Women's Experiences in Leadership in K-16 Science Education Communities, Becoming and Being

Katherine C. Wieseman · Molly H. Weinburgh
Editors

Women's Experiences in Leadership in K-16 Science Education Communities, Becoming and Being

Springer

Editors
Katherine C. Wieseman
Independent Educational Consultant
Formerly Western State College
of Colorado
Teacher Education Program
Gunnison CO 81231
USA
kwieseman@gmail.com

Molly H. Weinburgh
Texas Christian University
College of Education
Fort Worth TX 76129
USA
m.weinburgh@tcu.edu

ISBN 978-90-481-2236-3 e-ISBN 978-90-481-2239-4
DOI 10.1007/978-90-481-2239-4
Springer Dordrecht Heidelberg London New York

Library of Congress Control Number: 2009921808

Printed on acid-free paper

Springer is part of Springer Science+Business Media (www.springer.com)

Acknowledgments

On behalf of all the authors in the book, we would like to offer our deepest gratitude to the Association for Science Teacher Education (ASTE) and each individual named below who served as a reviewer for chapters within the book. The publication of this book would not have been possible without the support and endorsement of ASTE and the hard work of the ASTE Publications Committee. We offer a special thank you to Janice Koch and Mary Atwater, who have a lived wisdom that allows them to look at this book in unique and wonderful ways, for their contributions of the Opening and Closing of the book, respectively, and their continued leadership to us and colleagues in the profession. We also appreciate the knowledge and wisdom contributed by each reviewer of the chapters in the book, as well as feedback given by the doctoral students in science education at Texas Christian University. We would like to end by acknowledging all the people who have led each of us, yet have not been named nor mentioned in our chapters, but whose images and names we see in our mind's eye.

Jennifer Adams
Valerie Akerson
Ken Appleton
Phyllis Balcerzak
Cynthia J. Benton
Donna Berlin
Mark Bloom
Janet Briggs
Erica Brownstein
Lynn Bryan
Ruth Burkett
Brenda Capobianco
Janet Carlsen Powell
Nate Carnes
Ann Cavallo
Susan Chaplin
Marta Coleman
Mike Dias
Patricia M. Friedrichsen
James Gallagher
Leslie A. Gregg-Jolly
Mark Guy
Laura Henriques
Paul Haughton
Trish Hembree
Fran Huckaby
Leslie Jones
Paul Kuerbis
Mary McNeil
Angella Manhertz-Hartford
Wayne Melville
Cathy Milne
Hedy Moscovici
Linda V. Muir
Mike Padilla
Eileen Parsons
Janie Pollock
Rose Pringle
Sherrie Reynolds
Jessica Riccio
Patricia Richards
Mike Sakin
Susan Sciame-Giesecke
Anna Scott
Gail Shroyer
Jamila Simpson
Barbara Spector
Susan Stratton
Heather Thiessen-Reilly
Norm Thompson
William Thompson
Deborah Tippins
Meta Van Sickle
Emily Van Zee
Gary Varella
William Veal
Robert Wallace
Carol Wieseman
Cathy Yeotis

Contents

Author Biographic Descriptions

Mary M. Atwater is a science education professor and a social foundations adjunct professor at the University of Georgia. Her research focuses on the socio-cultural-political influences on science learning and teaching. She served ASTE in a variety of capacities such as Elections Committee Chair, Board of Directors member, and Publications Committee Chair and is actively involved and has served in leadership positions in NARST, NSTA, AMSE, AERA, and SSMA. She was named a 2008 AERA Inaugural Fellows and received the 2005 AMSE Legendary Award.

Bambi Bailey is an associate professor at University of Texas, Tyler. She has served on NSTA's Preservice Education Committee, reviewed for the National Science Teachers Association (NSTA) Special Program Area for the National Council for Accreditation of Teacher Education (NCATE), served as local co-chair for the Science Teachers' Association of Texas 2006 Science Conference, and sponsored a student chapter of NSTA.

Robin Bingham is Professor of Biology in the department of Natural and Environmental Sciences at Western State College in Gunnison, Colorado. Dr. Bingham is an evolutionary ecologist. Her recent research focuses on plant-insect interactions and the evolution of plant defenses against herbivores. Dr. Bingham's greatest joy as an educator is sharing her passion for basic research with undergraduates. She lives in Gunnison with her husband and daughter.

Diane C. Cudahy is an associate professor in the School of Education, Health and Human Performance at the College of Charleston, South Carolina. Her current assignments at the college include teaching foundations of education on the undergraduate and graduate level and directing the South Carolina Teaching Fellows Program. She served as the President of the School of Education, Health and Human Performance Advisory Board and associate dean. Her research interests are focused on the recruitment and retention of bright young men and women in the profession of teaching and the improvement of college teaching.

Kathleen S. Davis is an associate professor of science education at the University of Massachusetts Amherst. Her research focuses on issues of equity, teacher education, and policy and reform in science education and she has published in *JRST*, *Science Education*, and *JSTE*. Kathleen's NSF-funded projects include *STEM Connections*

(GK12) and *Science Education Online*, which were designed to engage elementary and middle school science teachers in inquiry-based and reform-based practices. She currently teaches graduate courses to preservice and inservice teachers. She began her teaching career as an elementary teacher in Illinois.

Wendy Files is a middle school science teacher in Baker, Oregon. She earned her B.S. in biology from Western State College of Colorado. Although she attended college in the east and Colorado and has taught in the southwest, she always finds her way back to the Pacific Northwest. Additionally, she tutors students around the country via the internet. She describes herself as a life-long learner.

Pamela Fraser-Abder is an associate professor and Director of Science Education at New York University. In 2008 she received the Steinhardt Teaching Excellence Award for extraordinary dedication to the transmission of knowledge. Her research has focused on the recruitment, preparation, and retention of urban science teachers through the lens of gender and culture. She has worked nationally and internationally on areas of gender, culture, and professional development of science teachers and science teacher educators. She has served on the board of NARST, SSMA, ASTE, and ICRSME and is active in NSTA.

Claudette Giscombe, who received her doctorate in science education from the University of Massachusetts Amherst, has served nationally and internationally in teaching, research, and educational administration. She has thirty-five years of experience in teaching biological science and science education at the secondary and tertiary levels. She has conducted innovative research in social justice focusing on minority representation in science. Her research findings on African American women in science have been presented at national and local conferences.

Janice Koch, is a professor of science education at Hofstra University on Long Island, New York where she directs IDEAS – the Institute for the Development of Education in the Advanced Sciences. She has many publications in edited texts and professional journals addressing science education, gender issues in the classroom, and making science accessible to underrepresented groups. She is the author of *Science Stories: Science methods for elementary and middle school teachers* (4th edition, 2009) and the introduction to education textbook, *So You Want to be a Teacher? Teaching and learning in the 21st Century* (2009).

Beth Shiner Klein is an associate professor at SUNY Cortland in the Childhood/Early Childhood Education Department where she teaches elementary science methods, technology methods, and environmental studies. She is the graduate coordinator for the MSED in Childhood Education and co-directs a five-year Teacher/Leader Quality Partnership grant that provides partnership activities for elementary preservice teachers. She has served ASTE as a member of the Board of Directors and co-chair of the Environmental Education Forum. She is also a NCATE Board of Examiner, appointed by NAAEE.

Sara Lamar is a first year teacher and also works part-time in the guidance office of Gunnison High School in Gunnison, Colorado. She began her undergraduate studies at Keene State College in New Hampshire and completed a B.S. degree in biology at Western State College in Gunnison. She is an active member of the area search and rescue team, and gets outdoors as often as possible to hike, bike, telemark ski, raft, travel, and jog.

Cyndy S. Leard earned a Ph.D. in science education from the University of South Florida. She has worked as a professional scientist, K-16 educator, informal education instructor, and program facilitator. She has been involved in numerous initiatives to bring scientists and educators together in both formal and informal settings. Her research interests include studying science teacher educator Ph.D. programs and candidate characteristics, the use of teambuilding strategies to teach science content, and marine scientist and educator partnerships. She acts as a consultant for Future Vision Productions.

Roxanne Greitz Miller is an assistant professor of teacher education at Chapman University in Orange, California. In addition to teaching graduate courses, Roxanne is an active researcher with over $3 million in grant funding over the last five years. She is a frequent journal contributor and presenter at national and international conferences, and is a Contributing Editor to NSTA's *Science Scope*. Her interests include the role of language and literacy in the construction of science knowledge, transference of science inquiry skills to other subject areas, and teacher professional development.

Hedy Moscovici is a professor in the Division of Teacher Education and the Director of the Center for Science Teacher Education at California State University, Dominguez Hills. Drawing from qualitative and socio/cultural methodologies she conducts research in urban settings where she studies issues related to the dynamics of power relations in science classrooms, influence of social and cultural discontinuity and the learning of science, intersection between critical pedagogy and leadership, and follows personal growth using self-studies. She has served and serves as Principal Investigator and Co-PI on numerous grants.

Tamara Holmlund Nelson is associate professor of science education in the Department of Teaching and Learning at Washington State University, Vancouver. Her research focuses on the transformation of science teachers' knowledge and beliefs through professional development (including preservice education) and how teachers translate their new understandings into classroom practice. Currently, as PI on an NSF-funded research grant, she is studying collaborative inquiry amongst secondary science and mathematics teachers.

Irene Osisioma, an assistant professor of science education in the Division of Teacher Education, California State University, Dominguez Hills, focuses her research on urban students' science achievement as it relates to socio-cultural issues, teacher beliefs, and science inquiry implementation. She is Co-PI for the NSF funded System Wide Change for All Learners and Educator's project, project director for the California Department of Education's California Math and Science Partnership project and the Nigerian Quality Based Education Development

Initiative. She is also a science educator for the Quality Educator's Development project.

Lisa Pike is a first grade teacher with over seventeen years of teaching experience. She has served as the Response to Intervention and Mentor/Induction Coordinator for the Gunnison Watershed School District. As part of her work on her Master of Arts Degree at the University of Colorado (2000–2002), she conducted, published, and presented research internationally on collaborative mentoring and preservice teacher education. She has also taught graduate level courses in cooperative teaching, mentoring, reflective practice, building literacy across the curriculum, and use research based strategies to impact student achievement.

Patricia R. Simpson is a professor of science education in the biology department at St. Cloud State University in Minnesota. She has served on the board of directors for ASTE and MNSTA, and numerous state and national committees regarding policy, assessment, and teacher licensure. She directed a Minnesota statewide university reform initiative on science and mathematics teacher education, and has been PI on numerous grants including most recently a statewide multi-university collaborative for online science licensure. She received the Outstanding Science Educator Award in 1999 from ASTE.

Barbara S. Spector, professor of science education at the University of South Florida, developed 3 major centers, received 61 grants totaling more than $7.5 million, and authored 505 publications, presentations, and media. She received multiple awards, including AAAS Fellow, was consultant to state and local education agencies, professional organizations, and business industry; served on advisory boards and steering committees for NSF and NASA; was Director of Education for National Sea Grant and the Marine Technology Society; was on the Boards of Directors for ASTE, NSTA, and IASTS; and chaired a variety of committees and task forces.

Meta Van Sickle is a professor and department chair for teacher education at the College of Charleston. She has served as program director for the M.Ed. program for science and mathematics teachers, and worked on many committees with the School of Education, been an active member of ASTE, NARST, and NSTA. She has served ASTE as a member of the Board of Directors where she earned the Presidential Citation for Outstanding Service, as a committee member, and chaired the Publications Committee. Her research interests include studying the ethics of care in science education and science for all including the linguistically diverse.

Starlin D. Weaver is an associate professor at Salisbury University in Maryland. She teaches courses in science pedagogy, classroom management, classroom assessment, and environmental education. She has served as an NSTA program reviewer for NCATE, a Board of Examiner member for NCATE appointed by AACTE, and the NCATE coordinator at her university. Her research interests include integration of environmental education into methodology courses, integration of reading and writing in the content area into secondary science methods courses and use of technology in the teaching of science and environmental education.

Molly H. Weinburgh is the William & Betty Adams Chair of Education and the Director of the Andrews Institute of Mathematics, Science, & Technology Education at Texas Christian University. She directs the Ph.D. in science education, teaches courses in science education, and teaches the Honors section of a non-majors biology course. Her service to science education includes: President of the ASTE, ASTE Board of Directors, Strand Coordinator of Culture, Social and Gender Issues of the NARST, and Policy Committee of the SSMA. Her research interests include equity issues, inquiry science, and professional development for teachers.

Katherine C. Wieseman has been a professor of teacher education at Western State College of Colorado. She served ASTE as conference co-chair and WISE Forum liaison, and is involved in Project Kaleidoscope. She promotes professional development focused on teacher change, inquiry learning, and ELLs' academic language development. Her scholarship encompasses web camera use, mentoring using interactive video conferencing, electronic portfolios, learning agreements as tools for empowerment, and collaborative mentoring. Whilst a Ph.D. student, she was in a study of leadership in science education and on NARST's International Committee.

Catherine G. Yeotis, emeritus faculty at Wichita State University, taught courses in science education, environmental education, and technology integration and facilitated the master's in C&I program for 28 years. She received funding for 31 grants that promoted inquiry teaching and learning in science classrooms. She is a past president of the Kansas Association of Teachers of Science, past director of SW-ASTE, Past President of ASTE, and served 2 years on the NSTA Board. She was a reviewer for the *JSTE* and the *JESE*. She received the Outstanding Science Educator Award from KATS in 1989, and the Outstanding Mentor Award from ASTE in 2004.

Opening

Everybody counts
When the spider
Weaves the Web
No connecting point
Is missed.
If you are missing
From our midst
We are the lesser
For that loss,
And incomplete.

~ *Brown, 2007*

Finding a home in the academy, settling in and attaining tenure, moving on to leadership and making your voice heard in science education is a lot like the process of weaving a web. It takes deliberate action, thoughtful determination, and careful consideration of how to connect the threads. You need to pace yourself, and build slowly toward your goal. This monograph can help. It provides insight into the ways in which women have surveyed the landscape of professional science education and navigated the challenges that are often unique to the intersections of leadership and gender.

Katherine Wieseman and Molly Weinburgh have collected and meticulously edited the stories of exceptional women in science education, whose desires to impact the field through their work and leadership skills, were often met with confounding experiences as they sought to embrace, what Judy Brown (2007) referred to as the feminine dimension of leadership – in their organizations, relationships with colleagues, students, science teachers, and indeed, within themselves.

What characterizes these stories is that they are specific and highly contextualized. They provide authentic descriptions of the ways women in science education integrated the lessons of their lives; their relationships with mentors, colleagues, students, and teachers; and the stories, poems and research that they read and to which they contributed in order to allow their lives to speak on behalf of their journey to leadership in science education.

What is significant are the ways these stories collect around the notion of the feminine, not as a gender, but as a construct, a quality in all of us. The range of qualities that speak to the feminine orientation in science education leadership include, but are not limited to, the use of stories to illustrate a point; the capacity for partnership and collaboration; and the ability to be accepting, appreciating, listening, and nurturing. The metaphor of the germinating seed in the organization of this monograph reminds us of how the solitary seed begins to grow and needs specific care and a garden in which it can thrive.

Whether introduced by men or by women, feminine approaches to leadership are often confronted by the cultural norms of traditional leadership models. Understanding how to navigate this territory while recognizing as legitimate, their feminine leadership qualities, these women's stories have much to teach us. Good leaders make other people better and the need for leaders to actively reflect on their practice is as crucial as the reflective practice in which we engage our future science teachers.

As the monograph begins, we visit the life stories of science education faculty and we notice that it is easy for women to be dazzled by the brilliance of their colleagues, stunned by the politics they encounter, and overwhelmed by the intellectual path they are following, with all its challenges and rewards. It is important to understand the loneliness experienced by these women as they document the journey toward the doctorate and professorial positions in science education. How inspirational to understand the critical role of mentors and how, in fact, an organization of women striving towards finding and hearing their voices in science education provided a structure of mentorship for so many others. So often, a secure group of colleagues is the default mode for men pursuing careers in professional science education; not so, for women. How important it is then that the Association for Science Teacher Education's forum, Women In Science Education (WISE), continues today and spawned this text.

Section two of the text further helps us to understand the struggles, conflicts, and ultimate gains made by women: some who confronted the demands and limitations of working within teaching universities; all who discovered the importance of being reflective about one's strengths and goals; and those who gained confidence by understanding the local context and taking reasonable risks on behalf of students, programs and their own development as science education leaders. Further, this section highlights the interventions on behalf of their own growth that women in science education implemented in order to bypass the negative constraints of institutions and programs and reach beyond the barriers presented by local contexts to make a difference to future science teachers. They accomplished this in part by being vigilant about recognizing their personal strengths and convictions and seeking mentors in the science education community. Ultimately these women created alternative pathways for their roles in science education and became significant leaders. They offer tools and advice for us to consider as we move forward in science education and an important example of the power of passion and self-discovery.

After the seed germinates, the complex plant emerges, representing intersections of culture and identity, as considerations of ethnicity and gender converge in section three of the text. These authors present ways to negotiate the layers of otherness that situate them as immigrants to the dominant culture of science education. These tools are useful to all of us – not only by example, but by inference. By that I mean the resources these authors share are of practical as well as symbolic use. They remind us that we are all part of what we want to change – that hierarchies need to be deconstructed – but not replaced by other, alternative hierarchies. Rather, the paradigm that replaces traditional hierarchies needs to be based on shared meaning, constructive collaboration, active listening, and meaningful dialogue. In this way, we are consistent learners, understanding the culture of others and becoming richer for that knowledge in the process.

The flowers that emerge in section four of the text provide models of leadership rooted in research, mentoring, active listening, and shared communities of learners and leaders. These authors remind us of the diverse ways that science education leadership presents itself and the common denominator for these contributions is in the domain of communication and collaboration, an aspect of feminine leadership, not limited to women's experience, but brilliantly informed by it. They remind us, as have other contributors to the text, that domains of leadership include the professional development domain, essential to promoting leadership in schools all over the country. What better way to promote science education leadership than to model it for our practicing teachers?

Finally, the plant returns to seed again and the voices of women with young careers in science education give us hope and optimism as we look forward to transformational leadership models in the science education community. This text itself is a model of what it seeks to promote. It opens spaces for dialogue by promoting shifts from knowing to questioning, extending a welcome to the readers, speaking from authentic experience, and bringing 100 percent of themselves to the conversation about science education leadership. The threads of the web connect intractably and the reader will be better able to handle the ups and downs of professional science education as a result of the voices of the authors and the editors. ....A heartfelt thanks to all the contributors and especially to Katherine and Molly for ensuring that their stories are shared.

Past-President of the Association
for Science Teacher Education
Fall 2008

Janice Koch

Reference

Brown, J. (2007). *A leader's guide to reflective practice.* Victoria, BC: Trafford Publishing.

Chapter 1
Introduction: Seeds Germinating and Growing

Katherine C. Wieseman and Molly H. Weinburgh

This book focuses on leadership with an emphasis on women as leaders in science education. It responds to the call in the Association for Science Teacher Education's *ASTE Newsletter* by then-President Simmons to "schedule regular professional development on leadership that targets leadership skills and capacities" for "all of us to develop, strengthen, and share leadership capacity in all contexts of our professional lives" (2005, p. 2). This book is critically needed because of the paucity of literature about women's leadership within the science education community. According to studies in science and mathematics education, leadership in these areas is not well understood (Gallagher et al., 2008a, 2008b). Leadership in science education may be different from leadership in other professional fields such as business and government, which are more widely studied. Gallagher et al. (2008a, 2008b) reported that factors such as research, writing, active membership in professional organizations and committee work lend to the acknowledgement of individuals as leaders who hold no formal leadership title. Furthermore they indicated that while doctoral programs are resources for leadership development, faculty members at well-recognized institutions of science education in the United States varied in their perceptions of the extent to which leadership development was explicitly targeted as an aspect of doctoral study experience.

Project Kaleidoscope (PKAL), an informal alliance advocating the transformation and improvement of undergraduate programs in the fields of science, technology, engineering, and mathematics (STEM), however, is focusing its efforts on leadership capacity development. This organization shifted its discourse to this focus in 2003 (PKAL, 1989–2008, 2004, 2007) in response to an

> awareness of the need for leaders – at all career stages and in all spheres of responsibility – to be working more intently and intentionally toward a common vision of what works, what matters, and what lasts in strengthening the undergraduate learning environment (PKAL, 1989–2008, para. 1).

K.C. Wieseman (✉)
Independent Educational Consultant, Formerly Western State College of Colorado, Gunnison, CO, USA
e-mail: kwieseman@gmail.com

K.C. Wieseman, M.H. Weinburgh (eds.), *Women's Experiences in Leadership in K-16 Science Education Communities, Becoming and Being*, DOI 10.1007/978-90-481-2239-4_1,

The work of PKAL, like this book, uses storytelling as a vehicle for sharing and disseminating knowledge and wisdom about leadership issues.

The content of this book represents courageous steps by contributing women authors to expose their innermost tensions, dilemmas, and feelings about leadership, making them available to others who may be living in similar circumstances. These women as leaders and authors are like gardeners planting seeds, which are full of potential and promise for growth. For this reason, we (Katherine and Molly) use the metaphor of a plant life cycle. The flowering plant life cycle metaphor captured our attention as editors because leadership as seen through the eyes of the authors, like a flowering plant's life, is about interdependence and growth/development of organisms in their environment; it is organic, cyclic, and never-ending. The metaphor allows illumination of leadership issues and representation of the journey of generating this book, constantly spotlighting the pervasiveness of leadership as personal and social activity, while explicating and honoring the myriad dimensions of leadership as distinct. We envision the chapters as helping to nurture conditions for germination and growth of leadership in individuals. To this end we

- outline the purpose of the book,
- identify potential audiences,
- share a story about the book's origin,
- present the theoretical foundation shaping the book as a whole, and
- preview the book's organization and content.

The *purpose* of this book is to provide inspiration and guidance for educators/academics in K-16 science education, as well as individuals in other professions, as they develop as leaders. The intent is to share the authors' wisdom about leadership, which has been carefully cultivated over the years. Although the content draws from the lives and wisdom of K-16 female science educators/academics in the United States, we believe that the tensions, dilemmas, and feelings are universal. Theoretical and practical orientations embracing relational ways of knowing are used to ground authors' sharing, examination, and discussion of leadership. We believe that this orientation has potential for other professional communities, including educational institutions, scientific and research organizations and laboratories, and the private sector. Furthermore, our understanding of relational ways of knowing and a desire to honor authors' experiences led us to a conscious decision as editors to support differences in style across chapters and preserve the beauty of authors' experiences.

We envision the *audiences* of this book to be both collective and individual. The collectives may be situated in university institutions, K-12 communities, grant projects, informal learning settings, scientifically oriented organizations, or the private sector. As a member of a collective, you may be reading this book with a group who are discussing the chapters and debating authors' message. As an individual, you may be exploring or engaged in research related to the issues and ideas shared in the book. You may be a leader grappling with the issues and ideas shared in the book. The chapters offer insight for you to construct personal definitions of leadership and heighten awareness of how and why you lead.

The *origin of this book* can be traced back to the 1995 Association for Science Teacher Education's annual meeting. Catherine Yeotis' sense of aloneness at her institution and need for a supportive mentoring community led her to gather a small group of female colleagues in science education to support and learn from each other, a community they called Women in Science Education (WISE). The purpose of this group was, and still is, to (1) share and discuss issues faced by women in science education in a social context that embraced women's ways of knowing and communicating, (2) mentor one another through various and changing life/professional phases, and (3) provide opportunities to collaborate on research interests. The publication of this book is a credit to the role which the WISE group (now the WISE Forum) in the ASTE has played in many women's professional lives.

Discussions during WISE receptions in which participants shared their struggles and dilemmas about professional issues united a small group of WISE Forum female members to collaborate in conference panel sessions at the annual meetings of the ASTE and the National Association for Research in Science Teaching (NARST). These sessions focused on dilemmas women were experiencing as they balanced personal and professional lives through crucial intersections of their lives (Wieseman et al., 2005a, 2005b). The storied format encouraged all session participants who had experienced the issues to join an emotionally honest discussion. The personal and academic lenses through which the stories were told and heard contributed to creating a socio-cultural context in which participants and co-facilitators could relate and resonate. Participants' requests for continued conference sessions as well as co-facilitators' desires to continue and further the discourse led to a WISE Forum-initiated workshop at the 2006 ASTE annual meeting in Portland, Oregon (Wieseman et al., 2006). A workshop on leadership development evolved from the realization that the balancing act dilemmas presented in the earlier conference sessions were because the women wanted to be leaders in science education. The workshop highlighted precursor activities that have led to the leadership activities in this book. In response to feedback indicating the need for a permanent record, session/workshop facilitators gathered around a table in the hotel lobby at the 2006 ASTE annual meeting to discuss their visions for responding to this feedback. Should it be a series of articles in a special issue of one of the science education journals? Should it be a book? What should the content be? For whom would this publication be intended? As organizer of this visioning meeting, Katherine quietly wrote notes, posed clarification questions, and offered her synthesis of the discussion to ensure that each contribution was heard in the light it was offered. Who should coordinate this writing endeavor? The women at this meeting looked to her – a logical choice. In this moment Katherine became immersed in an internal battle of confidence. Though fearful of the magnitude of such a formidable, foreign academic endeavor she agreed to organize it. Simultaneously she knew she would immediately seek the counsel of a knowledgeable and wise female elder in the international science education community – a woman who had become one of her mentors and one of the women involved in the collaborative project – Janice Koch. In a walk along the Willamette River, modeling servant leadership, Janice offered reassurance to

Katherine's crisis of confidence to undertake organizing and editing the book. Using a deliberately analytical approach with a criteria list, they discussed who Katherine could approach as a potential coeditor. The criteria included personal attributes such as who was timely and well organized, could easily organize and work with others, and was a good team player who could see big picture and small details; the temporal factor of whom would have time; and political considerations such as for whom it could be strategic for career advancement to be involved in coediting a book. Katherine asked Tamara Nelson to coedit the book. Tamara agreed but academic norms and politics of career advancement necessitated that she redirect her energies in the late summer of 2006, causing a need to identify a new coeditor. Katherine remembered all the women who had been co-facilitators of the conference sessions and workshop and slept on this dilemma, awakening one morning in December of 2006 with the insight, "How could I not have thought of Molly Weinburgh?" A phone call of inquiry and invitation was quickly placed to Molly, who quickly accepted with the words, "I am greatly honored." Having participated in two of the three presentations and all the WISE socials, Molly understood the need to collect, analyze, and share women's experiences in leadership and to offer the gift of this book back to WISE and to a much broader audience. A general invitation was issued for authors at the WISE Forum reception in 2007.

Using telephone and interactive video conferencing technologies, we (Katherine and Molly) deepened our knowledge of each other and potential authors. A book proposal was submitted to the ASTE Publications Committee and Board of Directors in the summer of 2007, and their endorsement transplanted the writing endeavor into a new context – generating an actual book. From its inception in 2004, the collaboration resulting in this book has been a multidimensional and complex story of leadership and deliberately reflects extensive discussion amongst individuals at conference sessions and within WISE.

A review of *recent scholarship* suggests that leadership has been viewed from as many as sixty-five classification systems grounded in different epistemologies (Northouse, 1997); and psychological, social, and organizational theories, each with particular views of the purposes and accepted manifestations of leadership (Astin & Astin, 2000; Chrobot-Mason & Ruderman 2004; Heifetz, 1994; London, 2002; Northouse, 1997; Shaffer, 2000; Spillane, 2006). In the field of science education findings from a National Science Foundation-funded study at Michigan State University reinforced earlier research that the understanding of leadership and leadership development is limited. In science education what is unclear is the definition of leadership, dimensions of leadership, and factors influencing leadership (Ashmann, Anderson, & Gallagher, 2003; Gallagher et al., 2008a, 2008b).

Is leadership a commodity, a personality trait, an aspect of emotional intelligence, a behavioral act, a style, a set of skills, a tool for goal achievement, an exercise of power in a relationship, a group process, a synonym for management, or an activity? Is leadership something that can only be assigned or ascribed to a position, title, or rank; or is it something that can emerge as a result of individuals' perceptions of those with great influence? These labels have been used to conceptualize and define leadership (Chrobot-Mason & Ruderman, 2004; Heifetz, 1994;

London, 2002; Northouse, 1997; Spillane & Diamond, 2007), and thus imply who can and cannot be a leader. Traditionally, in the United States leadership has been regarded from a male Western majority group or hegemonic, hierarchical perspective (Chrobot-Mason & Ruderman, 2004; Smith, 1997) and has often involved the use of coercion to affect change (Northouse, 1997). In their report on leadership in higher education, Astin & Astin (2000) claimed that the concept of leadership has been given little attention in tertiary level education. Furthermore they elaborated two models typically coexist in higher education in the United States: one being hierarchical in which the command structure is headed by a leader with individuals at the lowermost level having relatively little or no power or autonomy and the other being individualistic in which the leaders are those faculty who have gained the most professional status and recognition. In such systems, leadership and power become proportional to the individual's position and status within the hierarchy and professional recognition.

Shifts in consciousness as well as an increasing complexity and pluralism within groups, organizations, and institutions, which reflect existing hegemony, are forcing (1) reassessment of the construct of leadership (Bowman, 1997; Chrobot-Mason & Ruderman 2004; Smith, 1997), (2) examination of the notion that authority is a necessary ingredient for leadership (Flower, 1995), and (3) a challenge of the supposition that leadership is equated with management – a concept and practice designed to produce stability, increase efficiency and effectiveness in an organization, maintain order, and create consistency (Northouse, 1997; Spillane & Diamond, 2007). Different perspectives hold that leadership can be adaptive, dynamic, complex, and uncertain social activity (Hesselbein & Cohen, 1999; Parks, 2005; Wheatley, 2006), vision-based and transformative (London, 2002), shared or distributed among the participants (Bowman, 1997; De Pree, 2004; Smith, 1997; Spillane & Diamond, 2007), and stewardship of resources motivated by a desire to contribute (Bowman, 1997; De Pree, 2004). In these views, leadership engages participants because each participant holds a personally valued professional identity, and he or she must decide to participate. Many 21st-century researchers on leadership now promote a more collaborative approach (Astin & Astin, 2000; Wheatley, 2006) than was promoted earlier. The chapter authors in this book have examined these differing ideas and orientations on leadership promoted in the literature. The authors have challenged, accepted, rejected, and/or modified the ideas or orientations in their presentation of the theoretical frameworks upon which their chapters are based.

Twenty-first century leadership is associated with numerous interdependent qualities, including personal attributes (innate and learned), interpersonal skills, ability to lead change, and focus on results and/or learning (Ashmann et al., 2003; Chrobot-Mason & Ruderman, 2004; Gallagher et al., 2008a, 2008b; London, 2002; Simmons, 2005). Integrity, trust, clarity, communication, free flow of ideas, risk taking, acceptance of possibility of failure, flexibility, diversity, horizontality in relationships, collective action, community, and mentoring have been identified as attributes for personal and organizational leadership capacity (George, 2005; Simmons, 2005). London (2002) described leading from a psychological perspective as a discovering process. Key to this discovering process are self-insight (knowing one's motivations, interests, abilities), self-regulation, and knowledge of organization and situational conditions as well as capabilities of others. In viewing leadership as a function of versatility, Kaiser and Kaplan (2000) indicated that good leaders have to be behaviorally complex.

Words like we, relational, connectedness, collaborative, transformational, inspirational, inclusionary, creative, compassionate, women's ways of knowing, intercultural voices and understanding, community, discovering, sharing, and mentoring begin to concretize ideas shared across chapters in this book. Leadership as enacted and storied by the contributing authors is multi-faceted, complex, and highly contextualized, yet several overarching theoretical assumptions frame the book:

- Leadership is dynamic, tangible, and pervasive in daily work. It takes place in a social context of relationships – pervasive personal and social professional activity manifest through myriad models, each model stemming from theories about how persons in groups should relate to and interact with each other to make decisions impacting the group.
- It is necessary to listen to, value, honor, and celebrate the voices of women in their journeys of becoming and being leaders in science education communities.
- A multiple mentors orientation within a community of colleagues, suggested by Jeruchim and Shapiro (1992) as appropriate for mentorship involving women, provides opportunity for nurturing leadership capacity and experience. As science educators and researchers, we need to recognize, mentor, and retain all available talent and views in K-16 science education communities.

The chapters espouse collaborative, transformational, adaptive, and servant models of leadership, placing high value on knowledge of oneself and human qualities such as being caring, and authentic and honest collegiality. Authors' becoming and being leadership orientations aim to redress experienced separation, fragmentation, and isolation that are both visible and invisible parts of the fabric of an educator's and academic's work.

The *organization* of this book deviates from the norm by having an opening by Janice Koch and a closing by Mary Atwater, two well-recognized women leaders in science education. Subsequent chapters are deliberately and explicitly organized into subsections based on a flowering plant life cycle metaphor. Each chapter is comprised of four components: story, theoretical frameworks, leadership activity, and leadership response. The story component of the chapter provides an anchor for readers, so there is a shared experience regarding some aspect of leadership. The theoretical frameworks situate the story within an academic and scholarly discourse. Authors offer readers guidance and/or concrete ways to delve into the aspects of leadership addressed in the chapter's leadership activities. The activities emerge from the story and fit with the theoretical frameworks. Author(s) offer another layer of reflection on the aspects of leadership addressed in the chapter in the leadership response. The response is their attempt to synthesize, provide closure, offer wonderment, and pose new questions for readers to consider. In some chapters these components are explicit; in other chapters they are intermingled and implicit.

Authors' choices for presentation reflect their way of knowing, leadership style and experience, and relationship with writing as expression of experience.

Each author's experiences fit within distinct phases of the flowering plant's life cycle. Wendy Files and Sara Lamar are like new seeds following from the growth and maturation of early leaders like Catherine Yeotis, but with a difference of thirty years that changes the contexts in which these new seeds will germinate and grow. Like seeds, individuals need certain conditions for their leadership capacity to germinate and grow. Leadership can look like and be experienced in myriad models, like the diversity of flowers and seeds we notice in the natural world in which humans actively participate.

The subsections (*and contributing authors*) are

I. First Seeds: Aloneness Leads to Building Community (*C. Yeotis, B. Shiner Klein, and S. Weaver*)
II. Germinating and Growing: Dimensions of Leadership Development (*B. Bailey; B. Spector, P. Simpson, and C. Leard; R. Bingham*)
III. Transplanting and Growing: Physical-Geographic Relocation and Accompanying Huge Cultural Shifts (*H. Moscovici & I. Osisioma; P. Fraser-Abder; R. Miller*)
IV. Flowering and Seed Forming: Leadership Models (*C. Giscombe; K. Davis; M. Van Sickle & D. Cudahy; T. Nelson; L. Pike*)
V. New Seeds: Hopes and Visions for the Future (*W. Files; S. Lamar*)

In Part I: First Seeds, *Catherine Yeotis* shares how her sense of aloneness at her institution and need for professional community led to the birth of WISE. She and her coauthors *Beth Shiner Klein* and *Starlin Weaver* use autoethnography to share the power of mentoring and possibility from participating in an evolutionary learning community (Laszlo & Laszlo, 2000) for leadership development.

The authors of the chapters in Part II: Germinating and Growing focus on aspects of personal leadership capacity development. *Bambi Bailey* uses the lens of feminist transformational leadership to discuss the importance of knowing context, self, and others in leadership development at a teaching college. Her conscious decision to move from a research-intensive graduate program to a teaching-intensive faculty position resulted in changes in the way she developed her leadership capacity. *Barbara Spector, Patty Simpson*, and *Cyndy Leard* view leadership from a systems perspective with a focus on individual goals and roles. Using biographical critical episodes as a mechanism for exploring personal dispositions, they look at barriers (lemons) and times of change as times of opportunity and emerge as leaders (make lemonade). *Robin Bingham* shares how her passion and self-discovery as a scientist and science educator, as well as the power of women's connectedness, have contributed to becoming an academic leader in a science department in a small liberal arts college. She argues that although gender and internal authority matter, a woman can advance in her career, become a strong mentor of new faculty, and develop personally meaningful professional goals as a leader.

The authors of the chapters in Part III: Transplanting and Growing are individuals who have grappled with cultural tensions that can occur in leadership development when professional and family factors seem to collide as a result of decisions to immigrate to the United States or fragment a family in order to become, be, and grow in the world of science education. *Hedy Moscovici* and *Irene Osisioma* advance the idea of co-generative dialoguing and recognition of various sources of capital as a solution to address cultural discontinuity in developing a new culture and leadership roles within professional development projects. *Pamela Fraser-Abder* describes leadership in a Community of Practice, in which learning and culturally sensitive action emerge from knowledge construction through participation in specific, meaningful, professional activities in a cultural context. *Roxanne Miller* tantalizes the reader with situational leadership and transformational leadership perspectives when she shares her story of how relocation might or might not influence a science educator's leadership skills, position, and influence as a result of her decision not to become a trailing spouse.

The authors in Part IV: Flowering and Seed Forming examine different models of leadership. *Claudette Giscombe*'s view of researching as a form of leadership is a doorway for thinking about how risk taking, courage, personal, and social responsibility are aspects of leadership, particularly in the context of researching complex, tacit cultural, racial, and social diversity issues impacting teaching and learning from a nonhegemonic stance. *Kathleen Davis* describes her leadership efforts with preservice teachers, as she mentors them to make explicit preexisting knowledge, beliefs, and feelings so they can break barriers, move into the science education community, and see themselves as legitimate participants within this professional community. *Meta Van Sickle*'s and *Diane Cudahy*'s ethic of care and egalitarian model maximize shared governance and responsibility as a group seeks the best fit rather than the right answer, thus celebrating and encouraging diversity of thought within a collective by actively making space for multiple and diverse opinions. *Tamara Nelson* argues that in higher education her valued concept of community is rarely found, but can be developed through intentional energies, resulting in a community of inquiry with shared leadership employing norms of collaboration. *Lisa Pike* debunks the assumption that formal principalship is the only legitimate and valued leadership role in school communities by arguing that teacher leadership, mentoring, and listening to all voices lend to developing strong professional learning communities.

In Part V: New Seeds young women who are beginning their careers in science education share their beginning journey in leadership. *Wendy Files* examines the influence of self-efficacy, mentoring, and the development of learning communities in the lives of new teachers grappling with classroom leadership issues. *Sarah Lamar* considers risk taking, belongingness within a community, and knowing self to be essential aspects of leadership development and views leadership, like research, as a way to change thinking, solve problems, and move society forward. Their chapters bring us full circle in the life of leadership.

In the Epilogue of the book we, the coeditors, interweave our individual and collective journeys and personal definitions of leadership with a synthesis of the content of the book's chapters. One aspect of our leadership journeys is servant leadership in this book project – a relational journey that entailed organizing, coordinating, writing, editing, mentoring, and leading each other as well as contributing authors; and being mentored and led by contributing authors and reviewers. Finally, in the spirit of bringing the book full circle from opening to closing, the book ends with the wisdom of Mary Atwater, an established woman leader in the field of multicultural science education.

References

Ashmann, S., Anderson, C. W., & Gallagher, J. J. (2003). *The context for developing leadership for science education: An examination of the personal contexts of current leaders and recent Ph.D. graduates*. Paper presented at the international conference of the Association for the Education of Teachers of Science.

Astin, A. A., & Astin, H. S. (2000). *Leadership reconsidered: Engaging higher education in social change*. Battle Creek, MI: W.K. Kellogg Foundation.

Bowman, M. A. (1997). Popular approaches to leadership. In P. G. Northouse (Ed.), *Leadership: Theory and practice* (pp. 239–260). Thousand Oaks, CA: Sage.

Chrobot-Mason, D., & Ruderman, M. N. (2004). Leadership in a diverse workplace. In M. S. Stockdale & F. J. Crosby (Eds.), *The psychology and management of workplace diversity* (pp. 100–121). Malden, MA: Blackwell Publishing.

De Pree, M. (2004). *Leadership is an art*. New York: Currency Doubleday.

Flower, J. (1995). A conversation with Ronald Heifetz: Leadership without easy answers. *The Healthcare Forum Journal, 38*(4). Retrieved November 2, 2006, from http://www.well.com/~bbear/heifetz.html

Gallagher, J. J., Floden, R., Anderson, C., Ferrini-Mundy, J., Gwekwerere, Y., Ashmann, S. (2008a). *Executive summary. The context for developing leadership in mathematics and science education*. Unpublished paper, Michigan State University.

Gallagher, J. J., Floden, R., Anderson, C., Ferrini-Mundy, J., Gwekwerere, Y., & Ashmann, S. (2008b). *Project summary. The context for developing leadership in mathematics and science education*. Unpublished paper, Michigan State University.

George, M. D. (2005). What works – A PKAL essay. Personal characteristics of leaders. *Project Kaleidoscope, Volume IV: What works, what matters and what lasts*. Retrieved November 26, 2007, from http://www.pkal.org/documents/Vol4PersonalCharacteristics.cfm

Heifetz, R. (1994). *Leadership without easy answers*. Cambridge, MA: The Bleknap Press of Harvard University Press.

Hesselbein, F., & Cohen, P. M. (1999). *Leader to leader*. New York: The Drucker Foundation.

Jeruchim, J., & Shapiro, P. (1992). *Women, mentors, and success*. New York: Fawcett Columbine.

Kaiser, R. B., & Kaplan, R. E. (2000, April). *Getting at leadership versatility: The case of the forceful and enabling polarity*. Paper presented at the 15th annual conference of the Society for Industrial and Organizational Psychology, New Orleans.

Laszlo, K. C., & Laszlo, A. (2000). Learning to become: Creating evolutionary learning community through evolutionary systems design. *A coalition for self-learning*. Retrieved September 10, 2007, from http://www.creatinglearningcommunities.org/book/roots/laszlo2.htm

London, M. (2002).*Leadership development: Paths to self-insight and professional growth*. Mahwah, NJ: Lawrence Erlbaum Associates.

Northouse, P. G. (1997). *Leadership: Theory and practice*. Thousand Oaks, CA: Sage.

Parks, S. D. (2005). *Leadership can be taught: A bold approach for a complex world*. Boston, MA: Harvard Business School Press.

Project Kaleidoscope [PKAL]. (1989–2008). PKAL Phase V: 2003–2007. *History of PKAL.* Retrieved February 3, 2008, from http://www.pkal.org/documents/PKALPhaseV.cfm

PKAL (2004, July). *Leadership. Investing in the future. Building institutional leadership for natural science communities 2004–2006.* Retrieved November 26, 2007, from http://www.pkal.org/documents/leadership.pdf

PKAL (2007).What matters – A PKAL essay. Planning, leadership and community. *Project Kaleidoscope, Volume IV: What works, what matters and what lasts.* Retrieved November 26, 2007, from http://www.pkal.org/documents/Vol4PlanningLeadershipAndCommunity.cfm

Shaffer, J. C. (2000). *The leadership solution.* New York, NY: McGraw Hill.

Simmons, P. E. (2005). President's newsletter message. *ASTE Newsletter, 39*(3), 1–2.

Smith, D. M. (1997). Women and leadership. In P. G. Northouse (Ed.) *Leadership theory and practice* (pp. 204–238). Thousand Oaks, CA: Sage.

Spillane, J. P. (2006). *Distributed leadership.* San Francisco: Jossey-Bass.

Spillane, J. P., & Diamond, J. B. (2007). *Distributed leadership in practice.* New York: Teachers College Press.

Wheatley, M. J. (2006). *Leadership and the new science.* San Francisco: Berrett-Koehler Publishers.

Wieseman, K. C., Spector, B. S., Yeotis, C. G., Gilmer, P. J., Koch, J., Moscovici, H., et al. (2006, January). *Leadership development valuing women's ways of knowing: Personal capacity and leadership cultures valued institutionally and in professional science teacher education.* Workshop presented at the Association for Science Teacher Education International Conference, Portland, OR.

Wieseman, K. C., Yeotis, C., Weinburgh, M., Moscovici, H., Koch, J., Fraser-Abder, P., et al. (2005a, April). *Crucial intersections in academic/professional and personal lives: Women's experiences in science education.* Special colloquium presented at the annual meeting of the National Association for Research in Science Teaching, Dallas, TX.

Wieseman, K. C., Weinburgh, M., Moscovici, H., Fraser-Abder, P., Spector, B., Lew, L-Y., et al. (2005b, January). *Balancing academic/professional and personal lives: Women's experiences in science education.* Round table discussion presented at International Meeting of the Association for the Education of Teachers of Science, Colorado Springs, CO.

Part I
First Seeds: Aloneness Leads to Building Community

Chapter 2
Enculturation into Science Education: Comparing Pathways and Dilemmas

Catherine G. Yeotis, Beth Shiner Klein, and Starlin D. Weaver

Theoretical Framework

A review of the literature reveals numerous definitions of leadership and a variety of leadership styles. The styles range from hierarchical and position-based to egalitarian and people-oriented to task-oriented. Some leadership definitions emphasize position while others focus on relationships. We agree with Northouse (2004) that leadership is a process, unfolding over time, in which an individual influences a group of individuals to achieve a common goal. Leadership, as enacted by us, has its roots in three theoretical frameworks: transformative learning (Mezirow, 1991), mentoring (Ragins & Cotton, 1991); and collaboration (Trinidad & Normore, 2004). The leadership stories and activities we share in this chapter show evidence of a combination of these three frameworks.

Transformative Learning

Mezirow (1975, 1991, 2000) described transformative learning as a process that goes beyond the acquisition of factual knowledge in that it actually changes the knower in some meaningful way. As described in *Transformative dimensions of adult learning* (Mezirow, 1991), differences exist between learning during adulthood and learning during childhood. These differences appear to be important in understanding how and why individuals deal with new situations that are unlike any they have previously experienced. Particularly during adulthood, learning is equated with transformation and transformation results from making sense of disorienting dilemmas one experiences in life. A disorienting dilemma may be a life crisis, a major life transition, and/or the result of accumulated transformations in meaning-schemas over time (Mezirow, 1995), which serve as a catalyst for change. Not all new situations are disorienting dilemmas, but those that are require the individual

C.G. Yeotis (✉)
Wichita State University Emeritus Faculty, Wichita, KS, USA
e-mail: cathy.yeotis@wichita.edu

K.C. Wieseman, M.H. Weinburgh (eds.), *Women's Experiences in Leadership in K-16 Science Education Communities, Becoming and Being*, DOI 10.1007/978-90-481-2239-4_2,

to critically examine a new or unfamiliar event, reality, authority, or power. When adult learners critically reflect on and make sense of disorienting dilemmas based on past experiences, transformative learning occurs. It is very important to note that not all experiences are a source for transformation; only those that disrupt prior interpretations, lead to critical reflection, and prompt new meaning making, constitute transformative learning experiences.

DiBiase (2000) has written about the implications of Mezirow's theory of transformative learning for science educators. He suggested that knowing how to facilitate adult learning based on transformative theory is necessary for science education reform and creation of environments in which males and females feel comfortable and successful, and that leaders in science education must understand the need to change the thinking of adults (teachers and teacher educators).

The social context is an important component of transformative learning (or transformation) and it is not the same for men and women. Members of both genders may face disorienting dilemmas throughout their life experiences, but often the differences in the social contexts of career situations between males and females lead to different levels of transformative learning. It is important to provide each individual with the skills to deal with the various dilemmas he or she may face. Critical reflection can lead to positive solutions to disorienting dilemmas, empowerment within those situations, and possibly or ultimately leadership.

To better understand how women deal with disorienting dilemmas and critical reflection, Belenky, Clinchy, Goldberger, & Tarule (1986) used structured interviews with women to examine their ways of knowing. This research provoked much thought as it provided evidence that the needs of women in educational settings were not being met as well as those of their male counterparts.

Mentoring

Mentoring is usually described as a relationship in which one person provides support and wisdom to another person (Daloz, 1999). The need for a mentor may arise as the result of a disorienting dilemma. Ragins & Cotton (1991) explained that it is usually more difficult for women to find mentors in work situations. They further stated that women often perceive barriers, whether they are real or not, that were not perceived by men. Ragins' and Cotton's research revealed five perceived barriers cited by women. First, fewer women were in positions to act as mentors to other women. Therefore, women felt they had limited access to mentors who could help them in critically reflecting on a given situation. Second, women feared taking the initiative to find a good mentor. They did not want to appear too aggressive or to be rejected for being too pushy because they felt they were expected to be more passive. Third, men were not very willing to mentor women. They felt it would be too complicated and, therefore, easier for them to select and work with another man as a protégé, thus men had more opportunities to develop mentoring relationships in both formal and informal settings. Fourth, women cited a fear of disapproval by supervisors or coworkers concerning who they might select as a mentor. Fifth, women feared misinterpretation of the mentoring situation as a sexual advance.

In a review of the literature on gender and race in mentoring relationships, O'Neill (2002) found that in the past decade mentoring has received much attention. Some universities began addressing issues with programs like the Young Women Leaders Program at the Curry School of Education at the University of Virginia that pairs 7th grade girls with undergraduate women in a free year-long mentoring program to explore how women and girls can best reach their leadership potential. Other organizations such as Association for Women in Science and Women in Science and Technology began mentoring programs to model leadership styles used by women as well as foster leadership among girls and women.

In 1993 Shakeshaft stated that "the feminine leadership styles are not better or worse than the traditional male-oriented ones, they are just different" (cited in Trinidad & Normore, 2004, p. 105). Furthermore "the point of examining these differences is not to say one approach is right and one is wrong, but rather to help us understand that males and females may be coming from very different perspectives, and that unless we understand these differences, we are not likely to work well together" (p. 575). According to the research findings of Trinidad & Normore (2004), "...women adopt democratic and participative leadership styles in the corporate world and in education" (p. 574). In addition, the leadership styles that appear to be used by women develop through socialization processes that include building relationships, communication, consensus building, power as influence, and working together for a common purpose" (p. 574). When statistics show that women are less likely to receive tenure and promotion than men (Rapoport, 2004), one wonders if the leadership in their departments is male or female, and how that leadership promotes mentoring and collaboration for its faculty.

Collaboration

Working together in a structured, recursive manner in an intellectual process is known as collaboration. Typically, the individuals involved reach a common goal by sharing knowledge and building consensus (Bennis & Biederman, 1997). *Women's ways of knowing* (Belenky et al., 1986) helped provide a clearer understanding of how women work together and can lead others. The authors of this book coined the term "pajama party" sessions as their most productive ways to collaborate on their project. Meeting at homes with children running through, taking long walks to discuss the research, and a lot of informal discussions worked best for their style as women in higher education. Thus as will be noted in the stories later in this chapter, the informal nature of the group meetings can result in very positive and productive outcomes.

The studies by Belenky et al. (1986) and Trinidad & Normore (2004) provoked much thought as they provided evidence that the needs of women in educational settings were not being met as well as those of their male counterparts. Often women were discouraged from following career paths considered to be for men. How can females in these situations become empowered, believe in their abilities and authority to move ahead, be successful and demonstrate their leadership potential in their desired careers? As women reflect critically on what they know and how they learn, they realize they need as much mentoring and collaborative teamwork as men to be successful in the work they do and the leadership responsibilities they assume. Discouragement from others becomes a disorienting dilemma for many women who expect to receive the same advantages as their male counterparts.

Because their experiences growing up did not include this support, women often do not see collaboration and mentoring as apparent resources (Belenky et al., 1986). Yet Trinidad & Normore (2004) concluded that mentoring and collaboration are essential for the development of leadership skills in both men and women.

As we started looking back on our enculturation into academia and the decisions we have made, we felt like qualitative researchers writing an autoethnography study. Reed-Danahay (1997) described autoethnography as a genre of writing and research that connects personal to the cultural, placing the self within a social context. Critical personal reflection – vital in analyzing the experiences, including disorienting dilemmas – is an aspect of writing autoethnography. This analysis forces the writer to acknowledge and face the many uncomfortable situations encountered on the journey. In the stories we share in this chapter, we describe key events in our development, showing how these events and their social context have molded us as science educators and allowed us to become leaders in both the cause of science education and the support of other women in this field. The key events in the first story, written by a science educator who has reached emeritus status in her career, are disorienting dilemmas that led her to transform her thinking and learning, professional work, and leadership. This story is followed by the leadership activity – a group called Women In Science Education (WISE), which was her solution to resolving the many disorienting dilemmas she experienced. The second and third stories, written by science educators in the prime of their careers, are about the fruits of participating in this leadership activity. Their stories point out the power of mentoring and collaboration as foundations for leadership development and offer a response to the WISE leadership activity.

Cathy's Story of Learning and Leadership

Inspiration and Support

I consider this phase of my life a strong foundation on which my career and leadership in science education was built. I believe it started in the back room of my father's drug store where he had set up a makeshift chemistry set for my sister, brother, and me. Although we found great joy in working with the set, it was not something my female friends enjoyed doing with us. Women from my generation who followed a science career path were in the minority, and usually had a very strong or unique support system. I believe my ethnic background, my immediate and extended family support with their high regard for education, and my knowledge of their belief in the equality of the sexes provided me with the confidence I needed in order to work toward accomplishing my goals.

My father emigrated from Greece at the age of 18 and settled in Michigan. My mother's parents had come from Greece in 1905 and lived in Boston. For both families, their strong belief in the power of education was a guiding force. My father learned English, attended a prep school, matriculated to the University of Michigan, and graduated with a degree in pharmacy. My mother was accepted to Radcliffe, but could not attend when the depression hit and she became the sole support of her parents because her father had lost his business in the depression. Fortunately her parents thought that education for a woman was critical, and they insisted that she take all the college prep courses and all the business courses offered at her high school. This allowed her to secure an accounting position with salary enough to help pay the bills. My future college fund was started the day I was born as were my sister's and brother's. In this social and cultural context, no matter my gender, I was assured I could become whatever I wanted. This definitely established my strong support system and my inspiration.

The Devaluing the Education of Females: A Disorienting Dilemma

Once out in the real world I discovered that everyone didn't agree with my family's philosophy. Many of my friends said their parents didn't think they needed to go to college, because they were just going to get married and raise children. Of the ones who were going to college, most were looking towards becoming elementary school teachers, because they would have the summers off, they liked kids, and they would not need to know much more than their students.

In my own elementary education I ran across similar views. I will never forget on the first day of school in the fourth grade, the teacher asked us each to tell her what we wanted to be when we grow up. I said I wanted to be a pharmacist. She responded, "Oh you want to be a farmer's wife, how nice" (1951). Although that got my dander up, it did not dawn on me at the time that she did not understand the idea of gender equity and would never consider that a girl would want to go into a science-oriented field.

In high school about 50 percent of the students in the science and mathematics courses were female, so I didn't think I was any different than any of my friends. However, when decisions were being made about what we were going to do in college, I had counselors asking, "Why would you want to go into a science field?" (1957). Just the tone of their voices communicated that they did not think it was an appropriate field for a woman. Sure enough, when I took science courses in college, they were male dominated with all male instructors. That is when I began to feel a little like an outsider; I was not one of the guys. At this point I began to reflect critically on the discrepancies I was confronting. Was my family misleading me? Were they setting me up for failure? Would I run into too many barriers? Or, were my friends, counselors, and even teachers misguided in their thinking?

The dilemmas I faced led to the critical reflection one would face in any counter-intuitive event. This might be akin to what Paulo Freire describes as the process of conscientization, by which adults "achieve a deepening awareness of both the social/cultural reality which shapes their lives and... their capacity to transform that reality through action upon it" (Dover, 2007, p. 1). This dilemma for me helped me take action and proceed on my original path for an education in a science field.

Meanwhile, I was active in a very large Greek American organization with a different social/cultural context from my academic context. After being president of my local chapter and head of the state organization, I became the national president of the young Greek American women's section of this group. As part of my responsibilities I was sent on an excursion to Greece with the presidents of the men's, women's, and young men's group. This excursion included audiences with the Greece's Prime Minister, King and Queen, and the Patriarch of the Greek Orthodox Church. I include this because after taking on such a leadership role, it was difficult to go back to feeling like a second-class citizen in science education.

Bumping into Male Positions of Power: Another Disorienting Dilemma

When I interviewed for my first teaching job in the spring of 1963, the administrators as much as told me I had the high school biology job. We even discussed my ideas for developing a nature study area in the field behind the school. At that time Michigan State University was on the quarter system, I completed my degree in March and had decided to travel to Greece and Turkey during April and May. Therefore, I was asked to come back the first of June to sign the contract. Needless to say, I was very excited about the prospects of this position. Upon my return, I was told that the head basketball coach had resigned, so they needed to hire a biology teacher who could coach men's basketball. I didn't quite fit the bill; here was another jolt of reality.

When I started my first teaching position, I found myself in a male dominated department in a school district that tended to reward male teachers with perks more than the female teachers. I soon discovered that I had to be more assertive and let administrators know what I was doing in my classroom. I was not accustomed to tooting my own horn, but found it was the only way to be recognized. When the school district was going to send some teachers to the National Science Teachers Association (NSTA) conference in Washington DC, I immediately applied to go. Since I was the only female on the trip, among the group of eight teachers, instead of letting me enjoy the perk of a single room, the science coordinator made it a point to tell me my room cost as much as the double rooms for the men.

More Bumps with Male Positional Power: An Accumulation of Disorienting Dilemmas

Although I always felt I could make a difference in the 7th–12th grade classrooms, I discovered I could have an effect beyond that. When my district sent me to learn about new curriculum projects that the district was considering implementing, I suggested that the two of us who were sent to the National Science Foundation workshop conduct inservice sessions for the other teachers in the district. The other individual was not interested, so I took it on by myself. It was so exciting to have these teachers come back each week and share how the ideas generated in the previous sessions worked in their classrooms. This was the first time I considered going to graduate school for more than just to renew my teaching certificate.

From 1975 to 1978 while in a doctoral program at a big ten university, I worked closely with three male science education professors. Although they were very helpful, I didn't have other graduate students in my science education program with whom to share, discuss, or collaborate. Because the male professors' perspectives and concerns were very different than mine, I felt a void. It was difficult for me to move from a leadership role as the chair of a public school's science department with eight faculty members in a mixed gender environment to the role of a student in a male dominated department. This called for a different mind set from that to which I was accustomed.

As a graduate teaching assistant who was assigned to teach two sections of the elementary science methods course, I was required to watch some of the other (male) instructors teach it first, so that I would know what to do, and use true inquiry. The first thing I learned was that they all talked the talk of inquiry, but still liked being on center stage. Since the three male instructors were good friends, socialized together, and had been teaching these courses for a few years, I didn't feel I could comment on this, but I went ahead and facilitated my sections the way I thought inquiry should occur, mainly as a guide.

As I progressed through the program I selected my dissertation committee. To cover all my bases, I included a statistician and a scientist as well as the three science educators. Of course they were all male. As I gave them sections of my dissertation draft to read, it was interesting to hear their side comments like, "My wife did my editing" or "My wife typed mine for me." As a single person doing it on my own, I didn't think I needed to hear that. Again, I survived.

Finally Having a Male Supporter: A New Disorienting Dilemma

When I completed my degree there were only two science education positions advertised in the *Chronicle of Higher Education*, with hundreds of applicants for each. At this point I began questioning my career choice. Had I made a mistake when I gave up my teaching position to think I could reach more students by working with teachers? The situation was depressing, especially when male candidates ultimately were hired for both positions.

That spring at the NSTA meeting, I often encountered one of the faculty members from one of the two universities searching for a science educator trying to go to conference sessions. An entourage of assertive men was always following him. Most were trying to talk to him, to remind him they had applied for the position and maybe he would like to chat with them or have another copy of their vita to review. I watched as 10–12 men were following him, all trying to talk to him at the same time as he tried to ignore them. I chuckled knowing well that women would never approach him in this manner. One evening I was sitting with friends at a social function, and he sat down with us, explaining to the group that he was exhausted from all the candidates pestering him. I didn't say a word about the position nor that I had applied for it. Although I did not get that position, that fall I received a call from him asking if I would be willing to fill a visiting position for a faculty member who was going to be on sabbatical. He said I was his first choice, because the university and department wanted someone who could represent them out in the community and schools in a professional manner. This was one of the few times I felt encouraged that women's ways of approaching life were finally paying off.

Isolation in Academia: Another New Disorienting Dilemma

The social and cultural context of academia is like no other. In many university environments there is a specific protocol in the hierarchy that one is expected to follow. As a graduate student I felt I was working closely with faculty and understood what it would be like once I secured my own position in higher education. I quickly discovered how unaware I was of the intricacies of the culture of higher education. I discovered that I had not been privy to the inner workings of the department nor how decisions were being made. I found that being a science educator in a department of curriculum and instruction could be very lonely. When I did get a tenure track position, I found myself the only secondary science educator as well as the only science educator with a terminal degree and an interest in research. I didn't realize at the time how that would affect my future collegial relationships. Looking back, I believe these experiences influenced the paths I needed to take to be successful. Transformative learning was happening to me.

The need for collegiality, and a desire to work for the improvement of science education, led me to get more involved in professional science education organizations. As I did so, I found that there were many women from small universities at these conferences, often alone and/or without a cohort from their graduate days. We began to share concerns about how much service is enough, where to publish our research, where to find advice, and other typical professional questions. In doing so, we found that we were asking similar questions. We began to feel a unique support from one another, thus the idea to organize the Women in Science Education (WISE) Forum within Association for Science Teacher Education began to develop.

WISE as a Leadership Activity

A series of disorienting dilemmas experienced by women, each accompanied by critical reflection, led to transformative learning and a leadership style that reached out to other women for collaboration and mentoring is the leadership activity in this chapter. Women's participation and emergent leadership through the WISE Forum has also given rise to this monograph. You can also develop a group such as WISE in your setting. The primary focus of WISE was and is to promote mentoring and collaboration among its members. The WISE Forum gathers annually at the Association for Science Teacher Education (ASTE) conference for an evening event, where formal and informal mentoring occurs. Graduate students are recognized and faculty members are celebrated. WISE celebrates when new graduates are hired in science education positions and when current science educators earn tenure and promotion. Acknowledgements are made for research, grant funding, and/or service. New publications by members are announced. WISE also celebrates when its members marry, become mothers, or grandmothers. It is important to state that although the gatherings celebrate the accomplishments of women in science education, the group is open to both men and women. Additionally, WISE gatherings act as a clearinghouse for sharing ideas, discussing current issues, and setting up collaborations for future research and position statements.

The WISE Forum provides its members with awareness that they are not alone in their quest for success and leadership responsibilities in higher education, and allows them to become members of an extended learning community with like goals. Miller (2000), in his discussion of learning communities states, "...the community learning approach is rooted in the belief that there are many legitimate ways to learn, and many useful ways to support learning" (para. 1). We feel the WISE Forum is an appropriate vehicle for women who build on one another's knowledge in a friendly, supportive environment focusing on the way they learn best.

Laszlo and Laszlo (2000) described an exciting type of learning community called Evolutionary Learning Community (ELC).

> An emergent (self-designing) learning system demonstrating dynamic stability by adapting with its environment and generating developmental pathways that are sustainable in the context of broader evolutionary flows. ELC is a human activity system that strives toward sustainable pathways for evolutionary development in synergistic interaction with its milieu. It does so through individual and collective processes of empowerment and learning how to learn and through an ongoing commitment to evolutionary learning. "ELCs do not adapt their environment to their needs, nor do they simply adapt to their environment. Rather, they adapt with their environment in a dynamic of mutually sustaining evolutionary co-creation." Just as the concept "system" is more a pattern than a thing, ELC is best conceived as an ideal image of community that can serve as a beacon for the design of new social systems appropriate for a new evolutionary era. (para. 16)

This type of evolutionary learning community concept is ideal for the way women best learn and lead. WISE members work together to adapt to the environment in which they find themselves as science educators in higher education. As a forum, WISE has evolved to meet the present needs of its members, and it will continue to evolve to meet the future needs of its members.

WISE as a leadership activity has been successful in assisting women in their process toward tenure and promotion, thus positioning women with titles and ranks in formal leadership roles and positions. It has provided a national forum to advance valid ideas for research, teaching, and service. It has assisted women in science education to become aware of their own leadership styles, assisted them in understanding how to work positively with individuals whose leadership styles are different than their own, and provided them with access to a support system beyond their own institutions. It would be interesting to collect data on WISE learning communities that form and see how closely they resemble the ELC philosophy and structure. Is there more productivity from and leadership by women who become members of these types of learning communities than women who are not involved? Do women improve their productivity and leadership potential once they become involved in learning communities?

Leadership Response

The following two stories provide a response to WISE as a leadership activity. Both Starlin Weaver and Beth Shiner Klein had experienced some similar situations to Cathy prior to graduate school. They had supportive families that encouraged them to pursue their interests. They experienced disorienting dilemmas during their early careers as classroom teachers. Their stories differ from Cathy's when they entered graduate school and were introduced to ASTE and WISE.

Starlin's and Beth's stories of becoming and being young female leaders in science education show the impact of WISE as a leadership activity. These two science educators share what they consider to be the benefits of mentoring and collaboration opportunities they received through their participation in WISE. They further explain how these experiences have provided them with the skills and confidence to take on leadership roles in their respective fields. Their stories are two of many leadership responses, as evidenced that every author in this book has experienced the fruits of WISE as a leadership activity at one or more ASTE conferences.

Starlin's Story of Networking, Collaboration and Mentoring

During my first year of graduate school, 1995, I was introduced to ASTE, when I attended the annual meeting in Charleston, West Virginia. At this conference I connected with the WISE Forum, which I credit as being an important launch pad in helping me get my first position in academia, thus legitimizing my participation in higher education and my potential to move into the leadership roles. I remember my first WISE gathering as an intimate informal gathering that felt to me like a group of kindred spirits, where mentoring was occurring and collaborations were encouraged. This small group eventually evolved into a learning community for me. This gathering was pivotal in my career because I met Beth Klein. We began a friendship and professional relationship that has lasted for more than eleven years, helping us each become leaders locally, regionally and nationally.

Finding a job was helped by my association with the women in ASTE and WISE. Many ASTE members mentored me, offered wisdom and advice that I listened to and acted on. I felt confident as I entered the job market. Through many informal interactions with my female counterparts I felt empowered. Unlike Cathy, I was fortunate to be looking for a position at a time when there were many options. During two interviews it was obvious that I was being interviewed because I was a woman and they were seeking to increase the percentage of female faculty members at their institutions. Being a woman during my job search was beneficial, and I secured my current position at Salisbury University.

My story becomes intertwined with Beth's story because we discovered that we shared professional interests. We began a collaboration with another colleague in science education, Juanita Jo Matkins, in which we tried something very innovative. We used telecommunications technologies to have our elementary science methods students enter into their own collaboration to accomplish a task. We named our semester-long collaboration across three institutions of higher education the TEACH (Teaching Across Collaborative Highways) Project.

Being a relatively new science teacher educator at that time, I was learning the nuances of higher education. Like Cathy, I was the only full-time tenure-track science teacher educator at my university. However, WISE had helped me create a community in the form of the TEACH project that helped me overcome my sense of isolation. I looked forward to our interactions in which we supported and encouraged each other. Working collaboratively on a regular basis with Beth and Juanita Jo helped my leadership style to emerge. In addition, it added credibility to my teaching of elementary school science since my teaching experiences were all in a high school setting, and the scholarship emerging from it enhanced my vitae and helped me secure my tenure and promotion.

My leadership continued to grow with additional collaboration and scholarship in the area of environmental education within our preservice methods courses and the creation of literature-based science inquiry projects completed by our elementary methods candidates. We communicate on a regular basis and support one another in times of institutional turmoil offering mentoring to one another. This includes advice on issues of curriculum reform and revision, and just listening if the other one needs to vent or talk through a difficult problem. Technology has provided us with a quick tool for constant communication and we use it to support one another even though our institutions are 350 miles apart.

My professional relationship with Beth has provided me with leadership opportunities beyond my classroom and university. I joined Beth in attending the North East regional ASTE where I once again found the mentoring that I first found in WISE. Janice Koch, a fellow WISE member, and Marvin Druger made it a priority to provide support and opportunity for junior faculty who were pursuing tenure and promotion. They encouraged me to become the North East regional representative for ASTE Board. Since then I have been offered and taken many opportunities for leadership as defined by a position or a title. Each opportunity led to another but all started as a result of my involvement in ASTE and meeting Beth through WISE.

WISE has also been a vehicle for me to give back to my profession. I have assisted other female colleagues as they achieved tenure and promotion. I served as an external promotion and tenure reviewer for another WISE member. My most recent venture is coediting a monograph focusing on the inclusion of environmental education in science education courses. My coeditors in this venture include Beth and another ASTE colleague, Al Bodzin.

Beth's Story of Networking, Collaboration, and Mentoring

My doctoral coursework, completed in 1994, was taught primarily by male faculty. My committee was all male; however, I did not feel that was a disadvantage. The majority of male faculty with whom I worked seemed to be sensitized to women's educational issues. I had a wonderful supportive mentor as my major advisor, Preston Prather. Preston worked with me to customize my program to focus on gaining more content knowledge in integrated sciences and completing pedagogy coursework on curriculum, program evaluation, and the preparation of future elementary science teachers. He encouraged my leadership potential by offering me opportunities to supervise student teachers, teach graduate courses, conduct research, and helped me acquire my first publication. He knew that I needed some strong female science education role models, so he arranged funding for me to attend my first ASTE conference where he introduced me to the few women of leadership in the organization at the time. The most important event for me was being introduced to the president of the organization, Cathy Yeotis. Cathy was very welcoming and encouraging, and has continued to provide me with mentoring through the years. Two years after my first ASTE conference, she founded the WISE Forum, and I attended the first WISE meeting at the ASTE conference in Charleston, WV.

In the 1990s, I benefited from ASTE activities developed for female science educators by Cathy Yeotis and other women in leadership positions, both before and after WISE began. These ASTE activities included sessions on how to function in a higher education arena (that was still largely male dominated in the science and science education areas). These sessions gave advice on how to guard my time against extra service duties and other requests that are frequently asked of female faculty. Partly as a result of this support, I had the confidence to take on leadership roles, even as a doctoral student. Another science education doctoral student at my university and former teaching colleague, Juanita Jo Matkins, and I founded a local science teachers' organization and coordinated the first two yearly conferences for area teachers.

Once WISE was established, I felt I had a secure group of colleagues on whom I could call as I went through the next stages of my professional life, including the job search phase. From several offers I chose a position at a small liberal arts college because of a comfortable and genuine interview process, and a culture of integrated teaching and active learning models that matched my interests. The first few years were difficult however. I was finishing my dissertation while serving as not only the sole science educator, but also the only mathematics and technology educator. I had not had much, if any, preparation in these latter subject areas. I felt isolated having lost the close contact with science education colleagues during my doctoral work and scrambled to complete my dissertation by the institution imposed deadline while developing skills in mathematics and technology education. As I was the only female faculty member with environmental and technology backgrounds at the institution, I ended up on numerous college committees which took up much of my time.

Through ASTE and WISE, I had developed a support network. After meeting Starlin at the 1997 ASTE conference WISE function, she and I, along with colleague Juanita Jo Matkins, put together a risky endeavor of using cutting-edge collaborative technologies to co-teach elementary science methods courses at three different institutions. Each of us brought different skills, resources, and support to the project. Although using these technologies brought many challenges, the support network pulled us through.

Because of my institution's service and teaching demands, without this collaboration I would not have been able to pursue much in the way of scholarship. The outcomes of being in this small evolutionary learning community gave me additional credibility among my institutional colleagues, a stronger scholarly record and helped me secure grant funding for several projects. I believe this collaboration, along with several letters of support from ASTE and WISE colleagues, enabled me to be successful in gaining early tenure. Initially I was advised, by my male department chair, not to go up early for tenure because "Most of your scholarship is in collaboration with others, how do we know what you are capable of?" After he read my full application and saw the letters of support from colleagues at multiple institutions, he also gave his support.

Afterward I changed institutions, which brought new challenges, and new tenure and promotion processes to navigate. I also moved into another support and mentoring network, the Northeast Region of ASTE, some who are active WISE members, specifically the 2007 ASTE President, Janice Koch. Janice encouraged Star and me to take on leadership roles in the regional organization as well as at the national level.

I continue to pursue leadership roles through various positions at my institution and in professional organizations. I regularly attend WISE functions at each ASTE meeting and have begun to give back to WISE by providing some mentorship to junior faculty during ASTE meetings, trying to be welcoming and touching base with them each time I see them, as Cathy did (and continues to do) for me. I have also been mentoring a junior faculty member, bringing her into ASTE and WISE and now she has entered a leadership role by taking on the planning responsibility for WISE functions.

Conclusions

The disorienting dilemmas that many of us in the profession have experienced have served as catalysts for the changes that have been occurring in our professional organizations and our professional lives. Through the WISE Forum, members have engaged in transformative learning and found support to resolve these disorienting dilemmas. Support has helped members overcome barriers and gain the confidence to take on leadership roles. Additionally, members develop collaborations and find mentors within the group. O'Neill (2002) reported that in the past decade mentoring has received much attention. Coincidently, that was about the time that the WISE Forum came into existence. The stories by Beth, Starlin, and the authors in this monograph show the value of having a group such as WISE. The WISE learning community is providing an important forum for celebration of success and the emergence of leadership by women.

References

Belenky, M. F., Clinchy, B. M., Goldberger, N. R., & Tarule, J. M. (1986). *Women's ways of knowing: The development of self, voice and mind*. New York: Basic Books.

Bennis, W., & Biederman, P. (1997). *Organizing genius: The secrets of creative collaboration*. New York: Perseus Books.

Daloz, L. A. (1999). *Mentor: Guiding the journey of adult learners*. San Francisco: Jossey-Bass.

DiBiase, W. J. (2000). *Mezirow's theory of transformative learning with implications for science teacher educators*. (ERIC Document Reproduction Service No. ED452020)

Dover, K. H. (2007). Adult learning theorist Jack Mezirow. *Newsletter on adult education*. Retrieved March 31, 2007, from http://adulted.about.com/cs/learningtheory/a/mezirow.htm

Laszlo, K. C., & Laszlo, A. (2000). Learning to become: Creating evolutionary learning community through evolutionary systems design. *A coalition for self-learning*. Retrieved September 10, 2007, from http://www.creatinglearningcommunities.org/book/roots/laszlo2.htm

Mezirow, J. (1975). *Education for perspective transformation: Women's re-entry programs in community colleges*. New York: Teacher's College Press.

Mezirow, J. (1991). *Transformative dimensions of adult learning*. San Francisco: Jossey-Bass.

Mezirow, J. (1995). Transformation theory of adult learning. In M. R. Welton (Ed.), *In defense of the lifeworld* (pp. 39–70). New York: SUNY Press.

Mezirow, J. (2000). Learning to think like an adult: Core concepts of transformation theory. In J. Mezirow (Ed.), *Learning as transformation: Critical perspectives on a theory in progress* (pp. 3–33). San Francisco: Jossey-Bass.

Miller, R. (2000). Philosophies of learning communities. *A coalition for self-learning*. Retrieved September 10, 2007, from http://www.creatinglearningcommunities.org/book/roots/miller4.htm

Northouse, P. G. (2004). *Leadership: Theory and practice*. Thousand Oaks, CA: Sage.

O'Neill, R. M. (2002). Gender and race in mentoring relationships: A review of the literature. In D. Clutterbuck & B. R. Ragins (Eds.), *Mentoring and diversity* (pp. 1–24). Woburn, MA: Butterworth-Heinemann.

Ragins, B. R., & Cotton, J. (1991). Easier said than done: Gender differences in perceived barriers to gaining a mentor. *The Academy of Management Journal, 34*(4), 939–951.

Rapoport, A. I. NSF, Division of Science Resource Statistics. (2004). *Gender differences in the careers of academic scientists and engineers. NSF 04-323*. Arlington, VA: National Science Foundation.

Reed-Danahay, D. (1997). *Auto/ethnography*. New York: Berg.

Trinidad, C., & Normore, A. H. (2004). Leadership and gender: A dangerous liaison? *Leadership & Organization Development Journal, 26*(7), 574–590.

Part II
Germinating and Growing: Dimensions of Leadership Development

Chapter 3
Choosing a Teaching University: A Change in Identity?

Bambi Bailey

Many of you are doctoral students at research extensive institutions and will go on to further your careers at other research institutions. This chapter, however, is aimed at those of you who may choose a college or university that focuses on teaching and you wish to develop your leadership potential. It may also be helpful for individuals considering a career change to an institution that focuses on teaching. Despite the focus on science education, some of the strategies provided within the chapter may be useful to individuals considering a position in which they may be different from their peers, such as a person who focuses on applications of research working with a group oriented toward basic research.

How can you fulfill your goal of focusing on exemplary teaching of science content and/or pedagogy and remain active as a leader in the science education community? My purpose is to share how I have responded to this question in the context of selecting a college that focuses on teaching. I write in a style consistent with being a teacher/mentor. The voice you will read is the one I use when discussing the future with potential and beginning graduate students much like the "New Seeds" in the final chapters of this book. My story includes suggestions stemming from events based upon planning, trial and error, coincidence, and luck. Perhaps with the help of mentors and books like these, your leadership story will more closely fit your expectations of what leadership should look and feel like.

Theoretical Background

Leadership development models in the business world have been consistent regarding basic knowledge and skills necessary for progress toward leadership goals. Business leaders from Bennis (1989) to London (2002) agree that in order to become a leader within an institution you must (1) know yourself, (2) know your context, (3) know others, and (4) be capable of operating on instinct by deploying yourself and taking risks. Business models focus upon self-interest as requisite to

B. Bailey (✉)
School of Education, University of Texas, Tyler, TX, USA
e-mail: bambi_bailey@uttyler.edu

K.C. Wieseman, M.H. Weinburgh (eds.), *Women's Experiences in Leadership in K-16 Science Education Communities, Becoming and Being*, DOI 10.1007/978-90-481-2239-4_3,

Table 3.1 General business leadership development versus feminist transformational leadership development

Four aspects of leadership	Business goals	Feminist transformational goals
Knowing context	Efficiency, delegation, hierarchy, power relationships, leverage	Efficacy for all stakeholders, relationship development
Knowing self	Work from strengths, accommodate weaknesses, influence and assess own emotional triggers	Work from strengths to build community among the group; admit weaknesses so that others may lead in that area; assess and influence emotional, intellectual, and professional needs
Knowing others	Use strengths, develop efficient teams, develop appropriate incentives, assess and influence emotional triggers of employees	Mentor others to build strengths to better meet the goals of the community, develop relationships among stakeholders
Operating on instinct by deploying yourself and taking risks	Choose an acceptable level of risk for the context which will most likely lead to meeting goals	Providing leadership for change of gendered status quo that may prevent goals from being reached despite potential risks to self

leadership (London, 2002; Rao & Kelleher, 2000). Feminist scholars critique business leadership models for this reason and have developed a model based upon Burns' (1978) work called feminist transformational leadership. Feminist transformational leaders focus upon support for others in need; social justice, equity, and care (Antrobus, 2000; Burnier, 2003; Held, 1993; Noddings, 2002; Rao & Kelleher, 2000; Tronto, 1993). Despite the differences in basic philosophy, both models require similar beginning steps in leadership development albeit for differing reasons (see Table 3.1).

Based upon this basic knowledge for leadership development, one can analyze any employment situation and describe activities that facilitate leadership within an institution from either perspective. I am trying to use feminist transformational leadership to create a space for science teaching and to mentor students to act as leaders within their schools and the profession to teach for change.

The subsections of this chapter address basic knowledge requisite for leadership development. Within each subsection is a description of one aspect of basic knowledge, a brief comparison of the knowledge to my story, and leadership activities in the form of a discussion of methods one might use to develop each type of knowledge. I will focus upon developing leadership skills in science education (locally and broadly) rather than service and leadership focused at the college or university level because this is the path that I followed. I preface the discussion about the four aspects of leadership identified in the table above with a short description of requirements at teaching universities where I have worked to place in perspective later discussions about my story.

Teaching Universities

My first faculty position was within a College of Education at a very small Hispanic-serving institution in Texas. I taught one section of elementary mathematics and science teaching methods in one course. There were no secondary content methods courses – only a single general methods course. In addition, I taught a wide variety of other courses at the undergraduate and graduate levels. My travel allowance was \$600.00 per year. Faculty members were expected to be available in their offices from 8 to 5 daily for advising, tutoring students, and other student needs.

My second institution is somewhat larger. Faculty are much better funded for travel. The methods course for preservice elementary teachers includes overviews of mathematics, science, social studies, and technology methodologies. However, there is now a methods/content course that I am privileged to teach. Teaching the course does present challenges though because students intending to teach both elementary and secondary science take the same course. I am also teaching science content courses for those who are going to be elementary or middle school teachers. Similar to my previous institution, I still teach a variety of other courses within the College of Education. Faculty are also expected to be available to students from 8 to 5 and attend various student functions on Saturdays about once a month. Much like the student body in a small high school in which the most engaged students are involved in many extracurricular activities, faculty at teaching universities find themselves called on repeatedly for committee work, service projects, and overloads. Keep the requirements described in this paragraph in mind noting that a faculty member is expected to conduct research, and that the research has to fit within the context of the institution. Therefore, a research agenda must often fit within the confines of the teaching assignment. The following sections offer ideas and advice for how to develop and retain an identity as a science educator and develop leadership skills within the context and confines of one's teaching assignment.

Leadership Development Activities and Relationship to My Story

Knowing the Context

To begin to understand the context of a teaching university, you have to understand the following: (1) the mission of and pressures faced by the university; (2) expectations the university, college, and department have of you; (3) the colleague and student population with whom you work; and (4) your place within the context. Understanding the context in which you work lays a foundation upon which you may build your leadership skills within the institution and within your field.

In the past, teaching universities focused very much upon teaching, then service within the community, and finally research. In fact, some faculty members at teaching universities still tell prospective employees that a few presentations per year will satisfy the research requirements for tenure. Within some Colleges of Education teaching in professional development schools counts as scholarship as well. However, many teaching universities are now requiring more research publication and grant funding for tenure and promotion. This increased requirement is in addition to the higher teaching loads in a diversity of courses, service expectations, frequent need for faculty to teach overloads, and lack of graduate student assistance. This is a very blunt description of the requirements at some teaching universities, but keep in mind that because of these expectations one gains more interaction with students that can lead to more opportunities to impact their science teaching decisions. Generally, that is the reason people choose to teach at such an institution. I know that I chose a teaching university for this reason.

I chose a teaching university where I felt I could make a difference in science education. I believed that I could impact the science teaching of more elementary school teachers in a teaching university context than I could at any other type of institution. My goal was and is to help preservice teachers construct professional knowledge and scientific understanding so that science is not intimidating to the point of avoidance. My students and I work together to remediate their pedagogical content knowledge deficiencies which often stem from anxiety about science content, so that they, in turn, can meet the content needs of their students. For example, I offer help sessions for my content classes during which the students and I explore ways that will help them understand the material rather than just memorize it. We develop investigations that address knowledge deficiencies from the past that are preventing them from understanding the current content, use online simulation activities, and look up websites for children that can be used for additional help. I support students who are worried about feeling stupid by going with them to area science events that they would not have otherwise attended.

The very aspects of the context at teaching institutions that can inhibit research (i.e., having no graduate assistances; teaching a variety of courses outside of science education, like educational psychology; teaching a larger number of contact hours) also help further the goals my students and I set. I have learned to incorporate data collection and make a space for research within my working context. For example, I keep a journal in which I note student comments, strategies that seem to work better or worse than others, content that students find significantly more difficult, and aspects of pedagogy with which they are not familiar. From my journal jottings I can develop research studies such as one which focuses on the use of online simulations to enhance the learning of science content. Developing a working knowledge of my students and their needs was one aspect of knowing my context for which I was fully prepared. However, as I describe in the next paragraph, I made assumptions about another part of my context that negatively impacted the progress toward my goals.

The colleagues with whom you work are another part of your context. Perhaps you have a content specialist background, like I have in science education, and are working in a College of Education dominated by faculty whose background is elementary education and who have generalist teaching certification. These faculty may tend to avoid science just as much as elementary education students do. Naively, I believed that a content specialist would be welcome in this environment and that issues of science education would be left to my expertise. However, in my case, there was a general belief that if you know teaching strategies in general, you can teach others how to teach any content area. Shulman's pedagogical content knowledge (1986) was not a part of the education program in which I worked. Pedagogical knowledge refers to general theories, methodologies, and strategies of teaching. Content knowledge is defined as the deep understanding of content necessary to teach that content. Finally, pedagogical content knowledge focuses upon the particular strategies used to help students learn specific content. I firmly believe that pedagogical content knowledge is requisite to good teaching at any grade level. My colleagues believed that only pedagogical knowledge was the necessary focus in education courses, and that other colleges were solely responsible for the development of preservice teachers' content knowledge. After butting heads over issues of pedagogy in our methods courses, I realized that I did not understand that my presence and ideas specific to pedagogical content knowledge represented a denigration of my colleagues' original teacher training which generally focused on pedagogy and thus, their current teaching. I failed to understand the context in which I worked, and so I was not as effective for my students as I could have been, because at times I lapsed into frustration regarding teaching methods in addition to causing frustration for my colleagues. My identity as a science educator in a department of generalists who differently prioritized the knowledge for which the education department is responsible impacted my focus on my goals. My goals were for my students first to teach science and later to lead the science teaching at their schools. I had to step back and choose between pedagogical content knowledge and pedagogical knowledge. I chose science education and the preservice preparation of teachers of science as my first priority. My actions had to focus around, not the teaching in other courses, but finding ways to help my students learn to teach science. I chose to be a proud specialist in a college of proud generalists.

To retain an identity as a specialist; become a local, regional, or even national leader in science education; and survive in a world of generalists, you need to find ways to understand and reconcile the parts of your career. Several activities that may help you to learn about the larger context of your setting include: (a) reading; (b) interviewing; (c) journaling or keeping records of interactions; and (d) making adjustments to your teaching and research that will help to meet both your needs and the needs of the department and college (Lawson, 2006). Reading refers to reviewing the mission, goals, and objectives of your department, college and university before you read the tenure documents. Look at syllabi from the past and present, analyze general education and specific program requirements as well as course prerequisites that students must complete. Finally, read some of your colleagues' work. Their work can give you information that can serve as the basis for conversations that are a part of the next strategy for knowing context.

Talk to faculty about their goals for themselves and their students. Find areas of connection with your colleagues. Are there areas in which you can collaborate in both teaching and research? Are there areas that are sensitive within the department, college and/or university? Pose the same questions with regard to your administrators and your students. Keep track of what you learn through journaling or even simple charts, linking context to goals for teaching, research, and service.

Other methods of keeping track of what you learn about your context include power analysis diagrams, mind maps, and timelines of your plans. Power analysis diagrams are figures depicting power relationships within a system (Allen, 2003; Oshry, 1999; Value Based Management, 2007). They differ depending on the theoretical and philosophical base used to define the relationships. For example, in traditional business, power is a limited resource that some have and some do not (Allen, 2005). A power analysis diagram from the theoretical position of business would include individuals and/or entities with an analysis of the level of power exerted and the direction of power. On the other hand feminists dedicated to transformational leadership would define power as unlimited empowerment potential embodied in all (Allen, 2005; Held, 1993). In order to effect change, feminist transformational leaders help others find their power and use it to reach goals of equity and social justice. A power analysis diagram from this theoretical perspective would map the potential of individuals to make change. As you learn about the context in which you work, you may want to make a more traditional business power analysis diagram because that will represent the traditional structure of the university. Although your goals may bc transformational, you need to exist within this traditional business structure. In order to better understand your context, you may need to see both your goals and the existing power structure from the institution's viewpoint. Once you understand your context, you can develop transformational power analysis diagrams. Essentially what you are doing as you develop knowledge of context is gathering information so that you can meet the needs of your institution, the stakeholders, and yourself.

Another way to learn about the larger context of your setting involves making adjustment in your research and teaching that will help meet your needs and the needs of the department and college. Some ideas include developing integrated courses with colleagues, co-teaching courses, and conducting action research in classes. For example I worked with a technology education, science professor, and my students on action research regarding the efficacy of online simulations in my science content course. I used my teaching of content as a model for my students. My colleagues and my students appreciated that I wanted to improve my teaching and collaborate across campus, and I was able to develop presentations and publications from the results.

You can integrate your research with areas that include your colleague's expertise, participate in long distance research either through another university, a colleague, or a mentor that will bring positive attention to your university. It is wise to reflect on what you know about yourself, your colleagues, and your institution as you develop your research agenda. Finding the point(s) where the goals of the institution and your research meet is important and requires flexibility regarding the level of specificity of your research agenda. Developing a five-year plan may become a feat of flexibility, but the goals you choose will serve as a guide. You do have to be able to make adjustments to your plan without beating yourself up for failing to meet every goal because certain issues are out of your control. Make every effort to keep your goals within the realm of supporting your institution and the institution may be flexible with their demands as well. Your service activities normally support the goals of the institution anyway; and working with inservice teachers, preservice teachers, or science students can provide a setting for research.

As part of professional service you can further your leadership by volunteering for state and national committees, projects, and events. You benefit in two ways from such service: by building a reputation within the science education community, and by supporting your university. Building knowledge about your institution and relating your activities to this knowledge will help you meet your goals in science education, and open paths to leadership within the institution. I must warn you, however, that it may be difficult – if not impossible – to effectively serve both masters, the university and science education. If you develop strong leadership capabilities, at some point you may have to choose between leadership responsibilities within science education and more generalized leadership responsibilities at your teaching university. The opposite may also be true. As you develop leadership capabilities and become a change agent in science education, over time you may distance yourself from your university. Taking these issues into account as you make professional choices requires knowing yourself.

Knowing Yourself

To be able to interact effectively with your colleagues, students, administrators, and research collaborators you must know yourself, warts and all. If you do not know yourself, you may build yourself a little windowless closet of professional loneliness because you don't understand your own behavior, reactions to situations, or how your behaviors are interpreted by others. You cannot build a community of stakeholders to work toward goals if you do not know how you behave and how others interpret your behaviors (Bennis, 1989; London, 2002). It is also difficult to determine common goals if you do not know yourself.

You should be familiar with a number of aspects of your personality: general personality issues, academic values, leadership style, perceptions of others, humor, self-insight, self-management, and self/other rating tendencies (London, 2002). A variety of instruments and methods exist to learn more about these aspects of yourself. The Myers-Brigg Type Indicator can provide information about your personality in general and how your personality may influence interactions with people who have other personality types (The Myers and Brigg Foundation, n.d.). I know that I am an introvert, meaning that I derive energy in the absence of other people. If I am going to work for an extended time with group members who are extroverts, meaning they derive energy from the presence of people, I must make certain that I take at least one break to be alone so that I can remain energized for an entire meeting. Otherwise, the members of the group will be just gearing up as I begin slowing down.

A second way to learn about yourself is to ask others. You can ask for feedback on your performance and/or behaviors from colleagues, administrators, and students (Lawson, 2006; London, 2002). Sometimes you get feedback without asking.

Through one of my student's presentations and follow-up activities in a human diversity class I learned that (a) I was a geek, (b) my sense of humor reflected that fact, and (c) my students were okay with it so long as they knew they could think I was one without bothering me. It seems even my sense of humor is different than that of my colleagues and most of my students. However, embracing my inner geek and publicly acknowledging it helped students be comfortable with me much more quickly than taking myself too seriously. For that class it seemed important that another student called me a geek, and I simply laughed. They needed to know that it was acceptable to tell me what they saw about me that was different. It was imperative in that human diversity class that I develop a comfort level among students in order for the class to begin to work on more difficult topics. Otherwise, the course would have been a sham, an exercise to meet requirements and not a learning experience. Thus, by knowing and accepting myself, I was able to improve my teaching and my interactions with students, which led to more honest discussions about teaching science and helping others to teach science.

Consider checking with the career management and student leadership offices at your university for different types of instruments to help define your leadership strengths and weaknesses. Reading leadership books also can provide input especially if they provide simple surveys that allow you to learn about your leadership style and skills (Lawson, 2006). A third way to learn about yourself, particularly as a potential leader, is to take an assessment like the Life Styles Inventory (Human Synergistics International, 1989). This inventory asks you to fill out a version of the inventory as a self-assessment. You then send five copies of the inventory to colleagues in science education, to those who you think will respond positively, as well as to those you think will respond negatively. The results from colleagues are compiled into one graphic for you. You overlay their compiled results over the results of your self-assessment. The resulting graphic shows you the areas in which you and others agree and the areas in which you may have a blind spot either in a positive way or a negative way. Additional materials allow you to develop a growth plan. Interestingly, London (2002) found that people who tended to be good leaders tended to assess themselves at lower levels than others do on a variety of assessments because they assumed that everyone had the same skill level as they did, and so rated themselves as average. This tendency can account for blind spots in positive areas. In addition, London offered other self-assessment tools about career motivation, self-insight, and self-management. Other authors and websites (Hart & Waisman, 2005; Kaagan, 1999; Lawson, 2006; Mindtools Ltd., 1995) offer games for individuals and groups to use to develop leadership skills and get to know each other.

Completing the Life Styles Inventory made a significant difference in the way that I look at myself. Reading the results of the inventory encouraged me to begin thinking of myself as a potential leader and start to follow the guidelines I discuss here. As a result I became more involved in science education outside my university in order to meet my goals.

Knowing Others

Knowing others requires listening skills and the ability to get people to talk and focus on their answers. People are willing to provide information about themselves if you are sincerely listening and interested in their responses. Often people tell you things you need to know, but you may not be focusing on them. Random thoughts, lists of things to do, or formulating your reply to what they are saying keep you from hearing others and learning about them. By talking to your students and teachers, administrators and colleagues with whom you work in science education, you can find professional goals that you share with others.

Attending professional conferences regarding science teaching and science teacher education is a valuable way to learn about the issues in your area of interest at the local, state, and national level. Listening at these meetings and pausing before you speak may be your best tool in the beginning of your career. By doing so, you let people know by your actions that you take their ideas seriously. They are more likely to listen when you decide to propose taking actions that may involve risks.

I started an National Science Teachers Association (NSTA) student chapter because I began to really listen to the needs of our students who were seeking middle school science certification. These students take a variety of content courses, but do not have a major in science, so they do not have a content area cohort of colleagues. In addition, they take all their education courses with preservice elementary teachers. They have no point of contact with each other, no critical mass in their courses and no sense of connection to what is being taught in the way that the secondary and elementary majors do. As one student put it, they were the red-headed stepchildren of the college. Our first year of meetings focused on getting to know each other while sitting on the floor of my office, talking about teaching science, appreciating humor unique to middle school science (you know you teach science if you think it's funny to put lignite in your friend's/child's stocking at Christmas), and just comparing field experiences. Eventually we outgrew the office floor and went on to become a more typical organization. Recently, two of the founding members helped my local planning committee organize the state science teachers' conference. One preservice middle school science teacher led field trips and the other helped with volunteers. Thus, I was able to meet their immediate need for support and community with the NSTA student group which led to mentoring, valuing their professional activities as inservice teachers, and working together as colleagues in science education.

Deploying Yourself and Taking Risks

From a feminist transformational perspective leadership is equated with power as a relational and unlimited commodity (Rao & Kelleher, 2000) rather than power as an exclusionary and limited commodity. When you consider power from this perspective, the act of deploying yourself and taking risks means that you: (1) challenge the status quo, (2) define yourself as a force for transformation in your way of working for transformation to reach valued ends, (3) work with power as defined for transformation, (4) remain open to opposing thoughts and, (5) analyze power dynamics using visual organizers to better understand the context (Oshry, 1999; Rao & Kelleher, 2000). Power as a relational and unlimited commodity also requires that you search for areas of interaction within science education and your teaching, so your goals can enhance these areas of your career. You need to remain active at the state and national levels in science education. You must choose to accept responsibility to represent your university to the science education community in a way that reflects well on both the institution and you, even while you may need to challenge your institution and yourself. These behaviors are risky because they are leadership activities, and leadership itself is a risky endeavor (Bennis, 1989; London, 2002; Rao & Kelleher, 2000). Particularly if you have chosen to teach at a teaching university, developing as a leader in science education requires that you be diligent and constantly learn about context, yourself and your skills, and others, both inside and outside science education. Choosing to use the leadership skills you develop is the risky part, but it may be the only way to reach transformative goals.

Recently, the president of my institution made it known that he would retire in the next three years. I noted the intensified jockeying for position that was evident among the administration to one of my colleagues. At first she did not believe that people actually plan ahead to take advantage of a change or loss of leadership in an organization; but after taking time to watch, she began to notice the same kinds of behaviors I had. Of course, a change in president can make a huge difference to the faculty, particularly if the Faculty Senate is discussing whether the institution should be absorbed by a university system. If a faculty is not prepared to make their wishes known, they have no voice in the eventual selection. I began to seriously consider who needed to be on the Faculty Senate rather than simply go with the flow of whoever wanted a position. I paid more attention to the discussions about whether our institution was going to be absorbed by a university system and did not simply take the prevailing answer, "We will never join a system unless the state finds we have significant problems," as a given. Most importantly, I began listening to others as they chatted and learned more about what others saw in our future. All of this information is helping me build a traditional business power analysis diagram that connects the dots of scuttlebutt into groups jockeying for power. I am facing multifaceted issues that complicate analysis and make decisions based upon feminist transformational leadership difficult. By simply writing this chapter with its critique of the status quo at my institution, I risk disapproval. A university, or any other institution, is a system whose parts have a complicated relationship and approval from one part of the system can result in disapproval from another part of the system. Thus, knowing yourself and your goals allows you to look at the system and all of its parts to choose where your leadership will best serve the constituents and goals you support.

Leadership Response

The activities that I am recommending in this chapter for growing as a leader have served me well. This chapter and much the literature describe leadership development as though it is something that one completes after meeting certain professional positions. However, with every new context, problem, or colleague one must work through the aspects of leadership development to some extent. As you enter a situation you may find that one or more of the four aspects of leadership (knowing context, knowing self, knowing others, and taking risks) requires more energy and thought. Therefore, we are constantly becoming as well as being a leader. Leadership development may become second nature with experience, but one must continue to learn, problem solve and reflect upon the best way to meet one's leadership goals through the orientation within which one has chosen, in my case this is feminist transformational leadership.

As a final example about the continuing role and influence of the four aspects of leadership, and becoming and being a leader, I remember the situation of the firing of the science coordinator at the state level who advocated for multiple perspectives on the politically sensitive topic of teaching evolution in schools and also endorsed evolution over intelligent design as science curriculum. This event happened within the context of increased activity and power of the proponents of intelligent design within my state's education structure. Shortly after the firing of the science coordinator, I accepted the challenge of rewriting the state science standards which were due to be written according to our revision cycle. This leadership endeavor was particularly complex because we had lost a key player at the state level and confronted a strong pro-intelligent design contingency. To enact feminist transformational leadership in this environment was complicated. It required that I consider the impact of change on all stakeholders as part of an ethic of care (Antrobus, 2000; Burnier, 2003). I had to revisit the four aspects of leadership. I used what I knew about leadership to examine the context to (1) be a productive member of the task force, (2) learn what my colleagues' reactions were and could be, based on what I knew about them, and (3) assess what my preservice teachers needed to know and be able to do in order to teach science effectively in this political environment. One question that I asked myself was, how do I teach my preservice teachers to challenge the status quo while still embodying an ethic of care for them as individuals? I had to know myself as a science education professional and assess my own level of skill, knowledge, and courage in order to respond to the treat of intelligent design to science education. I had to be willing to take new risks. I also asked myself, how do I both meet the needs of my institution while challenging the state upon which the institution is dependent?

I cannot emphasize enough how important it is to realize that leadership development is not a process that reaches completion. We have to go through the process each time we face new, challenging issues in our professional lives.

References

Allen, A. (2005, October 19). Feminist perspectives on power. *Stanford encyclopedia of philosophy*. Retrieved October 31, 2007, from http://plato.stanford.edu/entries/feminist-power/#empo

Allen, J. (2003). *Lost geographies of power*. Oxford: Blackwell.

Antrobus, P. (2000). Transformational leadership: Advancing the agenda for gender justice. *Gender and Development, 8*(3), 50–56.

Bennis, W. (1989). *On becoming a leader*. Reading, MA: Addison-Wesley Publications.

Burnier, D. (2003). Other voices/Other rooms: Towards a care-centered public administration. *Administrative Theory and Praxis, 25*, 529–544.

Burns, J. M. (1978) *Leadership*. New York. Harper & Row.

Hart, L. B., & Waisman, C. S. (2005). *The leadership training activity book: 50 exercises for building effective leaders*. New York, NY: American Management Association.

Held, V. (1993). The ethics of care. In D. Copp (Ed.), *Oxford handbook of ethical theory* (pp. 1–44). New York: Oxford University Press.

Human Synergistics International. (1989). *Life styles inventory*. Plymouth, MI: Author.

Kaagan, S. S. (1999). *Leadership games: Experiential learning for organizational development*. Thousand Oaks, CA: Sage Publications.

Lawson, K. L. (2006). *Successful leadership skills*. London: Axis Publishing.

London, M. (2002). *Leadership development: Paths to self-insight and professional growth*. Mahwah, NJ: Lawrence Erlbaum Associates.

Mindtools, Ltd. (1995). Mind tools: Essential tools for excellent careers. Retrieved June 8, 2007, from http://www.mindtools.com/.

The Myers and Briggs Foundation. (n.d.). Type use for everyday life. Retrieved March 23, 2007, from http://www.myersbriggs.org/.

Noddings, N. (2002). *Starting at home: Caring and social policy*. Berkeley: University of California Press.

Oshry, B. (1999), *Leading systems: Lessons from the power lab*. San Francisco: Berrett-Koehler.

Rao, A., & Kelleher, D. (2000). Leadership for social transformation: Some ideas and questions on institutions and feminist leadership. *Gender and Development, 8*(3), 74–79.

Shulman, L. (1986, Spring). Those who understand: A conception of teacher knowledge. *American Educator, 10*(1), 9–15, 43–44.

Tronto, J. (1993). *Moral boundaries: A political argument for an ethic of care*. New York: Routledge.

Value Based Management. (n.d.). Force field analysis and diagram – Kurt Lewin. Retrieved November 5, 2007, from http://www.valuebasedmanagement.net/methods_lewin_force_field_analysis.html

Chapter 4
Making Lemonade from Lemons: A Road to Leadership for Women in Science Education

Barbara S. Spector, Patricia R. Simpson, and Cyndy Leard

We now recognize throughout our careers we had unknowingly been building on the wisdom of Naisbitt (2006) as stated in the following: "Times of change are times of opportunity. When relationships of people and things are shifting, new juxtapositions create new needs, and desires offer possibilities" (p. 92). Similar to the rest of society, the rate of change in higher education has accelerated. We see shifts in administration at all levels, faculty movement among institutions, national and state policy, university shrinking budgets, grant funding patterns, and subsequent colleagues' desires to preserve or expand their turf. Our responses to barriers emerging from these shifts were to let go of whatever was interfering with, or blocking our creativity and personal job satisfaction, and to seek new professional relationships with associated new tasks. This behavior invariably put us in leadership positions.

We understand leadership to be guiding, directing, or influencing people. Leadership opportunities may exist in various positions, settings, or roles. Leadership positions available in science education include project director, grant writer, department chair, manuscript author, officer in a professional organization, to name a few. Leadership settings exist in schools, institutions of higher education, government agencies, and professional organizations at the state, regional, or national level. Regardless of the setting or position, a leader needs to be able to diagnose a situation and shift roles as appropriate to achieve a desired goal.

Role shifting may occur at several levels within an organization. For instance, one may assume a role shift as one moves vertically in the hierarchy from the supervised to the supervisor. Another type of role shift occurs horizontally within an organization specifically during group or team product oriented processes. In these instances, members take on roles like truth seeker, direction setter, change agent, team builder, or spokesperson, to name a few.

We (women) have learned that our dispositions to create lemonade from lemons were identifiable and could be developed intentionally and deliberately. This chapter will facilitate other female science educators' entry to positions of leadership by

B.S. Spector (✉)
College of Education, University of South Florida, Tampa, FL, USA
e-mail: spector@coedu.usf.edu

K.C. Wieseman, M.H. Weinburgh (eds.), *Women's Experiences in Leadership in K-16 Science Education Communities, Becoming and Being*, DOI 10.1007/978-90-481-2239-4_4,

intentionally preparing them to shift roles when they experience changes in their circumstances.

Theoretical Framework

The theoretical basis for this chapter is multifaceted and is informed by literature from the following domains: organizational development, partnership development, and women's ways of knowing and working.

Organizational Development

Today organizations of all types, including higher education institutions, are considered "open systems" (Hatch, 1997; Senge, Kleiner, Roberts, Ross, & Smith, 1994). In such systems, organizations rather than being isolated from their environments act interdependently and exist in dynamic states of equilibrium (Harvey & Brown, 2001; Holzman, 1993; O'Neil, 1993; Senge et al., 1994). The organization constantly reacts to feedback from the environment in which it exists (Harvey & Brown, 2001). While the organization may appear to react as a whole, subsystems within the organization may respond in different ways. In order to understand this complexity of interaction, some organizational development theorists have described the entire system as a sociotechnical system that may be divided into a set of subsystems, each having its own purpose and focus within the organization. These subsystems include: structural, technical, psychosocial (cultural), goals and values, and the managerial subsystem (Harvey & Brown, 2001).

The suggestions made in this chapter assist in navigating through the culture of higher education to attain leadership. According to Harvey & Brown (2001), this cultural subsystem "includes the network of social relationships and behavioral patterns of members, such as norms, roles, and communications" (p. 40). It is not surprising then that our initial reaction to perceived roadblocks created realignment within this cultural subsystem or a seeking of alignment outside the immediate system, but within the network comprising the system and its environment. This realignment results in a restructuring of the relationship patterns of the people within the organization as they change positions, roles, or group identity.

Partnerships

If we move beyond our own institutions seeking new relationships, we must also be aware of the heightened importance of organizational culture as it impacts desired goals. For instance, as women, we may seek partnerships. Partnership, however, is an ambiguous term and may be synonymous with other words such as collaboration, cooperation, network, or alliance (Kingsley & O'Neil, 2004). We need to be vigilant as we filter these desired relationships through our perceptual screens. For some individuals partnerships may be casual, short-term events, and for others the term connotes a highly evolved, sustained relationship recognized by broad, in-depth communication and deeply rooted trust (Florian, 1999). In addition to the relationship issues that impact the development of partnerships, there are often funding issues that affect these relationships as well.

Women's Ways of Knowing

The work of Maltz, Borker, and Tannen support the important role of relationship seeking behavior in development of women as leaders in science education. Maltz & Borker (1982) observed boys and girls at play. They found that boys generally played in large hierarchically organized groups and preferred games involving elaborate sets of rules in which the end result was winners and losers. Girls, on the other hand, were more often found in small, intimate groups taking part in noncompetitive activities in which directions were spoken as suggestions rather than orders to be followed. Tannen (1994) in *Talking Nine to Five* seeks to explain how these behaviors play out in the work place. She suggests that men and women can be viewed as two separate cultures if a culture is defined as "a network of habits and patterns gleaned from past experience" (Tannen, 1986, p. 133).

Since men and women have different past experiences, they carry their unique cultures with them into the world of work. These experiences not only influence their preferred working styles and interactions of each culture, but also the ways in which they communicate both verbally and nonverbally. This suggests that as a result of being women, we follow women's preferred ways of working in response to perceived roadblocks by avoiding direct conflict. There is also research to support a biological basis for women to respond to adversity (perceived roadblocks) by gathering with other people especially women (Berkowitz, 2002). These women's ways of knowing and working lead to win–win situations common to partnerships, rather than win–lose situations common to hierarchies.

Our Stories as Critical Episodes

Women constantly encounter barriers to successful leadership in science education. The question emerges, what can we learn from the experiences of two women about their dispositions and behaviors that have led to leadership in science education?

Our stories served as the basis for the recommendations provided in this chapter. Over the years, we have shared numerous experiences and personal responses to those experiences. We compared our responses and then generated suggestions that can be used to help others interpret their own experiences. Each of us reflected upon and analyzed our career experiences and our responses to those experiences. We have shared these stories with each other over the past 20 years, and have provided support and suggestions to one another. In reviewing our conversations, we noticed patterns emerging from our life stories. We shared these patterns and suggestions with Cyndy, a relative newcomer to science education. She provided her perspective on our stories and related them to literature. She also helped us sort through our stories to identify those with meaning to the next generation.

The stories that follow reflect common types of barriers that emerged in our careers, and our responses to them. We have labeled these stories critical episodes. Each episode begins with a title identifying the writer, followed by a list of behaviors that are evident within the episode. These behaviors align with the recommendations list found at the end of this chapter. In addition, the first story sets the context for the study and introduces the main participants and their mentoring relationship.

Patty's Take a Chance

Behaviors emphasized: Seek mentors. Find alternative lenses. Use your knowledge and skills in new roles, positions, and settings.

It was my last semester in graduate school and my journey toward a Ph.D. in science education was almost over. As I prepared for my future, I submitted papers to regional science educator meetings in the hopes that my future employer would provide funding for at least one of these presentations. I also began the arduous task of finding a job. I approached this task in an organized manner searching the *Chronicle of Higher Education*, perusing the list serves of the major science education organizations, and keeping my ears open for mention of an opportunity that might arise during my daily routine.

It was a hectic time. I had completed the data collection and analysis for my dissertation and was writing up the results. I was teaching an elementary methods course and sitting in on a course in the Biology Department on endangered species while I wrote and rewrote sections of my dissertation. I had been offered another year of support at my university, but it was time to find a real job. My husband and I were in our third year of living apart and trips back and forth for weekend visits were getting old. After sifting through countless job advertisements and sending in numerous vitae with supporting documentation, I was invited to interview for several positions including a position in the biology department at a large, midwestern state university. With great anticipation and some fatigue, I secured a plane ticket, packed my bags, and flew off to the interview.

The first evening I had a chance at dinner to meet faculty from the biology department. They seemed very friendly but a little uncertain about what topics I might pursue for research. Their areas of research in genetics and microbiology were beyond my understanding, and my interest in teacher practice and curriculum were beyond theirs. The next morning's breakfast with the science education group was more comfortable. The group met weekly to talk about their students, programs, and what was happening in the state in regard to teacher education. They greeted me warmly and seemed pleased by my areas of interest and enthusiasm for teaching.

They had all been at the university for the majority of their careers and saw it as a good place to work. They told me that the union made it an equitable place to work, and the travel monies guaranteed by contract were shared by departments, so there was not a problem with finding travel funds. They told me there were people willing to collaborate on grant writing and research, and that I would love the senior faculty member who was currently on leave. The place seemed like a good fit. I could teach and continue my professional development through meetings and work on grants and publications with my colleagues. The chair of the search committee was especially supportive in promising to work with me in the latter two areas if I took the job.

The more I talked with the chair of the committee, the more I realized that I had a real opportunity to work in a place that fit with my vision of enhancing science teaching in the elementary classroom. I could teach science content in the way it should be taught and still teach methods at the secondary level. I could also share my STS experiences with local teachers through summer workshops. I also felt encouraged by the search committee since they insisted that I certainly would be successful in this environment.

In the weeks that followed, I anxiously awaited word from the university, and finally it arrived in the form of a phone call. Since I had also received another offer from a College of Education at about the same time, I had a hard decision to make. As I happily shared my news with my major professor, I was again shocked back into reality.

"You need to take the job in the College of Education," he said. "You need to be wary of science departments! Science educators cannot be successful and tenured in those departments! The philosophies and requirements are too different."

Not to be deterred, I made my decision to accept the position despite the advice of my professor. The encouragement of the search committee and the conversations that I had with the committee chair, who seemed very supportive and willing to act as my mentor, also made it seem like a good fit for me.

In August, however, as my husband and I finished packing the moving van to head north, I received a phone call from the committee chair, my future mentor. He called to let me know he was leaving the university to accept a position elsewhere. I was stunned. It was too late to change our decision; we had a house waiting for us and no other job options. Now I would be working most closely with a person whom I had not met. We continued to pack, and set off hundreds of miles away from friends, family, and professional colleagues. I arrived in Minnesota with anticipation. However, I was not prepared for the chilly reception that I received from my two closest colleagues. The reception from the first individual was understandable; this person had been my competition, finishing second to me in the final interview process. I understood the awkward position this put us both in as he was being asked to continue teaching classes without being offered a tenure earning position.

The second colleague was the senior faculty member who was on leave during my hiring process. He seemed friendly at first. Yet a few weeks into the semester I discovered he had been criticizing my teaching performance to others, because I was making changes in one of the courses he perceived to be his. This was just the first of a series of roadblocks he created during my first year, while telling me what a wonderful job I was doing and how successful I could be at the university just by teaching my courses and not worrying about publications or grant writing. He was not willing to share the summer teaching load, so I could earn extra salary. This left me in need of pursuing grants, something with which I had been promised help, including opportunities to collaborate. He was not interested in collaborating on any grants or research or writing. He preferred to work alone, or at least not with a female. He had already established consulting jobs and publishing work. In addition, he appeared to undermine my career by providing me with false information about requirements for work with schools, tenure, and promotion. He continued to provide me with misinformation throughout my career.

Meanwhile, the funding policies under which I had been hired changed significantly. Now travel monies were in short supply. Besides these professional obstacles, I was beginning to feel the pressure in my personal life as well. I felt isolated from my support system having left family and friends behind. It was hard to meet new people in the community because I spent so much time on work, and the women in the department were very heavily involved with their own teaching and research. I also felt the stress of being the sole financial support for my husband and myself, since he was in veterinary school. This pressure heightened as we realized that it was more expensive to live in Minnesota than we had anticipated. To top it off the cold, dreary winter months were fast approaching. These factors led me to reconsider my job choice as the semester progressed. I began to wonder if any of the jobs I had turned down were still available. In this frame of mind I took off for Tulsa, Oklahoma, and a regional NSTA meeting to make a presentation on my dissertation research.

"Thank goodness, I decided to submit those proposals," I thought to myself, as I boarded the plane, "I'm going to enjoy these few days in a new environment."

As I moved from session to session on the first day, I noticed several other people following the same session strands.

"Hello," I said to a tall, attractive, blonde woman, as I slid into the seat beside her.

From that point on, we hit it off exchanging names and brief biographies. We found we shared a common interest in STS education. We agreed to spend more time together over the remainder of the meeting. That is how Barbara and I first met. We talked until 2 AM the first night and for several hours over the next few days. During this time the conversation turned to the situation at my university. As I began to elaborate, Barbara would ask questions.

"I'm just feeling like I made a mistake," I said, "I don't feel like I can be successful in the science department under these circumstances."

"Why?" Barbara questioned, "Science departments are really good places to work. Look at all the resources they have for doing science. They have equipment, supplies, and support technicians. You won't have those in a college of education."

"Well, that's true, but I don't have anyone to work with," I complained, "I would like to write some grants, but I just don't have the experience to do it on my own."

"Well if your colleague does not want to write grants, that leaves the door wide open for you," she said, "You won't be perceived as being in his territory, because he's made it clear that he's not interested."

"We could work together on a project," she continued, "I'm always looking for a collaborator."

"I have an STS project I'm working on," she invited, "You also might want to join AETS, the Association for the Education of Teachers of Science (now ASTE). Many of the members are university science educators who are doing research and writing grants. You could find someone to work with outside of your university."

Through our conversations, she gave me a lot to think about and changed my perspective on my current situation. Instead of perceiving it as a hostile environment fraught with land mines just waiting to explode, I now realized I had before me a land of opportunity waiting to be explored.

As a result, here we are almost 20 years later still working together. I stayed at my institution and found it to be a good base from which to work. I achieved both early tenure and promotions. I have been involved in collaborations across and between departments, universities, and states. I have worked for state and national policy agencies and served in leadership roles in several science education organizations. For example, I was the director of Transforming Teacher Education, the higher education component of SciMathMN, a state funded systemic initiative to reform teacher preparation in Minnesota. I have worked on policy for CCSSO, AAAS, and NRC. I was director of the NCASTE and served for six years on the national board of ASTE. I also served two terms on the board for MnNSTA as well as other state and national organizations. I have passed on what Barbara helped me learn about leadership and opportunity to other science educators. I have even been able, on occasion, to help Barbara change the lenses through which she views her own circumstances.

Patty's You Have Done Enough

Behaviors emphasized: Find alternative lenses. Keep and balance your goals.

I was in the midst of tenure and promotion. I had successful student evaluations for my teaching. I had completed grants for summer teacher professional development programs. I had been making presentations at state, regional, and national meetings on a regular basis, and I had begun to be active at the state level in terms of committee work. Despite all this, I did not feel as if I was being respected by my peers in the department. I tried to figure out why. Was it because I was female in an overwhelmingly male department? Was it because they didn't understand and value my work in science education as opposed to science? Was it because I was not working hard enough/not teaching enough classes? Not having enough students? Surely I was doing something wrong.

Late one afternoon I stopped to talk with our secretary who had been there for years and seemed to have a good understanding of the inner-workings of the department and its faculty. I asked her what more I needed to do to meet the expectations of the department. She replied, "You don't understand. It is not that you need to do more. You are doing so much now that you make what they are doing look bad." Oh! That was certainly a new way to look at my life. I would never have come up with that answer on my own. My mindset of never being good enough had led me in entirely the wrong direction. The dean gave me a comparable message saying, "You are doing too much and will burn out if you do not slow down." I was not going to solve my problem, because I certainly was not going to stop doing things to attain my goals. At least I had insight into the situation and no longer needed to waste my energy trying to make them happy. It doesn't matter who hands you a new lens as long as you trust the person and he/she has an alternative view of the situation you are trying to reexamine.

Barbara's Navigating the Shifting Sands of Higher Education

Behaviors emphasized: Be collegial and build alliances to exploit opportunities. Use your knowledge and skills in new roles, positions, and settings. Find alternative lenses.

A two-year moratorium was imposed by the department chair on new entrants into the science education doctoral program, which I had led for many years, in order to build a cohort of adequate size to fill doctoral courses when offered. During the second year of the moratorium, family illness required me to take an extended leave from the university for the better part of a semester. Upon my return I found the department chair with whom I had negotiated my leave had left. An interim chair was appointed for two years. At the same time there was an interim dean appointed in our college, the governor disbanded the state Higher Education Board of Regents and appointed a local governing board for our university. That board refused to acknowledge the legality of the union. The new interim chair of my department refused to honor any of the former chair's commitments to me, including financial and assignment. She facilitated a junior colleague taking over the doctoral program and gave me a teaching assignment that did not provide access to the emerging doctoral cohort. There was no grievance procedure available, because the union had no status.

About this time I accepted an invitation to a community wide meeting addressing marine education convened by the Center for Ocean Sciences Education Excellence-Florida (COSEE-FL). There I learned scientists in the College of Marine Science (CMS) were in need of my expertise as a change agent and organizational development person. Since I wanted a doctoral level course to teach, I offered to share my expertise by teaching a course facilitating ocean scientists and educators working together to effect policy and change in science education. I was able to provide this course as part of my regular assignment by using an existing doctoral level course number and title. Science professors, post-doctoral scientists and engineers, and other stakeholders enrolled. Shortly thereafter, changes in personnel in CMS led to my becoming the Principle Investigator and Director of the COSEE-FL. This was one of seven National Science Foundation funded multi-million dollar centers comprising the National COSEE Network, whose mission was to "spark and nurture collaborations among scientists and educators to advance ocean discovery and make known the vital role of the ocean in our lives" (Sptizer & Ganter, 2007, p. 2).

Thus I moved into a national leadership position in ocean science education that included influencing national policy [e.g. with input to *An Ocean Blueprint for the 21st Century* (United States Commission on Ocean Policy, 2004)] and implementation of numerous innovations through this center. For example, I developed and pilot tested a sequence of five new doctoral courses related to community building among scientists, educators, and other stakeholders and a new model for constructing courses to meet the needs of specific audiences and constituents.

As an outgrowth of the alliances developed during the community building courses, stakeholders from informal science education institutions (free-choice education) informed me of the need to further educate themselves and their colleagues. In response, I moved into still another domain, with lots for me to learn, and a new audience for science education in my college. We developed and pilot tested four doctoral courses comprising an *Informal Science Institutions Environmental Education Certificate Program.* When one door closes, one searches for another that allows entrance and an alternative way to meet one's goals.

Barbara's Make the Boys Happy

Behaviors emphasized: Use your knowledge and skills in new roles, positions, and settings. Avoid direct competition.

When the reform movement came to the fore in Florida in the 1980s, I garnered several grants to develop new graduate courses modeling principles of reform. In concert with a mathematics education professor, we established nine new courses in which science and mathematics students learned together the *USF Graduate Program of Excellence for Mathematics and Science Teachers (GpEMSt).* This combined with lots of other grants resulted in a problem when I went up for promotion to full professor a year early. The department committee voted no. However my chair and the college wide committee over-rode their vote and said "yes" to my promotion. The dean, a golfing buddy of the three male science educators in my department, said "no." In a conference with my chair and the dean to inquire about what else was needed to qualify for full professor, the dean explained, "You have to make the boys happy."

"How am I supposed to do that?" Barbara asked. "You make them look bad because you are doing too much," said the dean.

Since the boys were not involved on the national scene, I responded, "OK, I will keep a low profile here and go back to working at the national level." I became Research Director for NSTA and a board member for AETS (now ASTE), a board member in NASTS (now IASTS), and traveled a lot the next year as a consultant to state education agencies and other organizations outside Florida. My promotion was granted the following year.

Patty's Environmental Education Partnership

Behaviors emphasized: Create a research/theory/practice cycle.

The teaching load at my institution is 12 credits each semester and since we are in a science department that often means more then 12 contact hours as labs do not get full credit. Besides teaching, there are requirements for service, scholarly work, and professional development. Without some sort of alignment or overlap of these four areas, faculty has little chance of meeting these requirements.

My dissertation research related to the impact of an instructional strategy designed to enhance environmental action behaviors in students. The strategy required students to engage in independent research related to people's knowledge, beliefs, and behaviors regarding environmental issues. Students then used what they had learned through the research to take personal action to positively impact the issue under investigation. The process taught students skills for inquiry and enhanced knowledge of the environment. By continuing to use this research as a central organizer, I was able to obtain grants for teacher professional development, use grant participants for further research leading to presentations and publications, and shape public policy to include environmental standards in current licensure proposals. This resulted in new courses and a need for our students to learn about inquiry through an environmental lens. I achieved early tenure and promotion through a cycle of research, grants, and practice and the satisfaction of teaching topics of interest to me.

Barbara's Making Enough Time

Behavior emphasized: Create a research/theory/practice/cycle.

I moved to a new institution after four years at my first university job. The provost asked for an exit interview with me when he heard I was leaving. He said "I have really been curious about you. Many people said you are just an unfocused opportunist writing grants for unrelated tasks every time an RFP comes out. I suspect you have a master plan for how these seemingly disparate pieces could come together."

I said, "You are right." I had been molding each proposal to fit into a process for systemic reform of science education that I envisioned when I was in Washington, D.C. as a grants program officer in the National Sea Grant College Program. Having been trained as a change agent in graduate school, I wanted to learn how change occurred in all facets of the education enterprise. Whether it was money for a children's summer camp, working with inservice teachers, revamping science courses, being a field test site for BSCS curriculum development, research on policy, and more, I wrote for it. Then I shaped the products or tasks into an articulated whole. That's how eventually the *GpEMSt* came into being with *World of Water* (*WOW*) summer camp as a testing ground for theories relating to change in teachers and students. The emergent theories from my research on *GpEMS*t and on *WOW* were implemented in schools that were the field test sites. I did research on the development and implementation of every grant product and used the emergent theories to develop and implement the next grant as a step in my vision of systemic reform. In this way, I was able to use my grant funds to create a coherent research-theory-practice and back to research cycle for my career.

Leadership Activities and Responses as Recommendations

Following are a set of recommendations explicated in the previous critical episodes that reiterate the behavioral strategies for success we have used to navigate the science education enterprise. In addition, we recommend you survey the landscape by using the inventory we have created. This inventory, located at the end of the chapter, can help you examine your environment for potential barriers, partners, alliances, skills, and opportunities. We include a final caution. Don't forget to delete. Remember to let go of what you no longer need or find of interest. Just because you are good at something does not mean it will bring you satisfaction. This final recommendation comes to us from Naisbitt (2006), and it is one neither of us does successfully. We all try to do too much. No one can do it all.

Seek Mentors

Find several mentors. Mentors can serve a variety of roles. They can give you a new lens, show you the ropes, provide a shoulder to lean on, serve as a new partner, provide needed support when things look bleak, and direct you to new connections.

Find Alternative Lenses

Remember to ask for another person's view of your world. Your perception of a barrier may actually be an opportunity. Choose individuals who can see your environment through different perspectives, such as home, school, or the nation.

Keep and Balance Your Goals

Keep your goals, both professional and personal, in the forefront of your decision-making. It is important to consciously balance your goals. For example, an opportunity may meet your current goal of tenure, but not meet your goal of having more family time.

Be Collegial and Build Alliances to Exploit Opportunities

Connect with people to achieve a goal, do a project, or fulfill a need. Seek others within your institution and outside your institution with whom you share common goals, compatible knowledge and skills, common dispositions, or necessary resources. For example, early career people need publications, so if this describes you, find someone to work with to achieve this goal. If you need resources to teach, then attaining grants might be the goal. If you want to teach a special course or type of course, then finding students might be the goal. If you want to do research, then obtaining respondents/participants might be the goal.

Use Your Knowledge and Skills in New Roles, Positions, and Settings

Be creative. Make notes from your inventory of all knowledge, skills, partners, venues, positions etc. Place each idea on a card and begin to group and regroup the items to see what opportunities might exist. Don't give up on your goal; just find a new way to achieve it. Look for new ways to use knowledge and skills to shift roles, settings and/or positions.

Avoid Direct Competition

Look for a variety of sources of rewards. Do not waste time and energy competing with someone else for the same prize. If you cannot get what you want, redirect your energy and focus to look for something else that is as good or better. In our profession, there are numerous opportunities for advancement, grants, publication, graduate students, and courses. If you cannot get the reward one way, pursue another way to get it. Explore multiple avenues through the university, professional organizations, or school districts. Often resources within a department, such as office space, students, and courses are limited. Therefore, you need to decide how important a particular reward is to you. For example, if you are competing for a particular course, investigate whether another existing course is just as good for you or whether you can create a new course, perhaps for a new audience, which will be equally rewarding.

Create a Research/Theory/Practice Cycle

Look for ways to integrate your practice with grant and research work. We are all asked to do more than we can reasonably achieve. You can start with any part of the cycle and advance to the others.

Final Thoughts

Many individuals, especially women in science teacher education are more inclined to avoid conflict with colleagues than most men. In order for teacher education to benefit from the strengths women science educators bring to the teaching profession, it is essential for women to learn to recognize and pursue opportunities for alternative pathways for success in academia. Changing a pathway does not suggest giving up on your goal; rather it is just avoiding the barriers that prevent you from achieving it. Barriers create new context. Take advantage of the new opportunities created by the new context. This allows women within the professoriate to align their ways of working with their female oriented relationship seeking behaviors rather than forcing them to abandon these collaborative methods for more androcentric behaviors. By retaining women in science education, the enterprise will continue to grow and benefit from the unique perspective that women bring to the science education professoriate.

Inventory

Women can intentionally and deliberately develop dispositions to create lemonade from lemons by identifying key cultural and political information. The following inventory creates a sensitizing screen to aid in the development of these dispositions and increases awareness of potential for successful options. The tool is reminiscent of the SWOT (Strengths, Weaknesses, Opportunities, Threats) analysis template found in organizational development literature (SWOT, 2006). The inventory is composed of two parts. Part 1 consists of a set of questions to guide you through an individual situational analysis. After you have closely examined your current situation and collected all the information you can about who, what, how, why, and when, then it is time to change your lens. Part 2 consists of a set of questions to help you reframe your thinking. Look at all the information from the perspective of not what problems exist in your life to be solved, but rather what opportunities exist for success.

The questions found in the inventory in Table 4.1 are designed to stimulate and guide self-reflection and analysis. The groups of questions overlap and are not exhaustive. They should act as a trigger for further inquiry. Their use can create self-awareness of leadership potential. Periodic review of personal environments can assist other women in accelerating their paths to leadership in a deliberate and comfortable fashion.

Table 4.1 Inventory

Part 1: Identify your current situation	
A. Goals ∗ What is your goal? (sensitizing question – personal versus professional) ∗ Can you identify mechanisms to achieve that goal? ∗ Does your personal satisfaction come from achievement, affiliation, or power?	**B. Barriers** ∗ What are the barriers to achieving your goals? ∗ Are they internal or external, personal or professional, or institutional? ∗ How can you mitigate, eliminate, or turn a barrier into an opportunity?
C. Roles ∗ What roles do you enact in your career and personal life that serve as barriers? ∗ Are these roles overt or covert?	**D. Knowledge and skills** ∗ What knowledge and skills do you have? ∗ Do you enjoy practicing the skills? ∗ What skills and knowledge do you want?
E. Competition ∗ Who is competing with you? Why? ∗ What are you competing for? ∗ Can you attain the prize in another way? ∗ Do you both need to win? ∗ Are there other prizes?	**F. Mentors** ∗ Who can provide you with another perspective? ∗ Why should you accept his/her view? ∗ Are there mentors in your personal life? ∗ What sources of mentors exist?
Part 2: Identify what is going on in your personal or professional worlds that can lead to opportunity?	
A. Relationships ∗ With whom can you collaborate? ∗ How do you connect with them? ∗ What can we work on? ∗ What strengths and goals do we share? ∗ What tasks must we both achieve? ∗ How are your skills complementary? ∗ How do you build successful relationships? ∗ What can we learn?	**B. Venues or tasks to attaining your goals** ∗ What are you willing to give up? ∗ How will this opportunity change life (enhance or diminish it)? ∗ How does an opportunity provide evidence for tenure and promotion? ∗ Does the opportunity help you achieve your present goal?

References

Berkowitz, G. (2002). UCLA study on friendship among women an alternative to fight or flight. *Current Psychiatry Reports, 4*, 441–448.

Florian, J. E. (1999). *Addressing evaluation of collaborations: Capacity building by school districts.* Paper presented at the Annual Meeting of the American Evaluation Association, Orlando, FL.

Harvey, D., & Brown, D. R. (2001). *An experiential approach to organization development* (6th ed.). Englewood Cliffs, NJ: Prentice Hall.

Hatch, M. J. (1997). *Organization theory: Modern symbolic and postmodern perspectives.* New York, NY: Oxford University Press, Inc.

Holzman, M. (1993). What is systemic change? *Educational Leadership, 51*(1), 18.

Kingsley, G., & O'Neil, D. V. (2004). *Performance measurement in public-private partnerships: Learning from praxis, constructing a conceptual model.* Paper presented at the American Society for Public Administration 65th National Conference. Portland, OR.

Maltz, D. N., & Borker, R. (1982). A cultural approach to male–female miscommunication. In J. J. Gumprez (Ed.), *Language and social identity* (pp. 196–216). Cambridge: Cambridge University Press.
Naisbitt, J. (2006). *Mind set!: Reset your thinking and see the future*. New York: HarperCollins.
O'Neil, J. (1993). Turning the system on its head. *Educational Leadership, 51*(1), 8–13.
Senge, P. M., Kleiner, A., Roberts, C., Ross, R. B., & Smith, B. J. (1994). *The fifth discipline: The art and practice of the learning organization.* New York, NY: Doubleday.
Sptizer, W., & Ganter, S. (2007). COSEE: The evolution of a national network and its strategic goals. *Current; The Journal of Marine Education, 51*(1), 2–3.
SWOT. *A Dictionary of Business and Management*. J. Law (Ed.). Oxford University Press, 2006. *Oxford Reference Online*. Oxford University Press. University of South Florida. Retrieved September 5, 2007, from http://www.oxfordreference.com/views/ENTRY.html?subview=Main&entry=t18.e6259
Tannen, D. (1994). *Talking 9 to 5.* New York, NY: William Morrow and Company, Inc.
Tannen, D. (1986). *That's not what I meant!* New York, NY: Ballantine Books.
U. S. Commission on Ocean Policy. (2004). *An ocean blueprint for the 21st century.* Final Report. Washington, D.C.

Chapter 5
Women's Ways of Leading: Navigating Gender Issues in the World of Science

Robin Bingham

Introduction

I am passionate about what I do. I am a scientist and an educator. In my mind, it is not incidental that I am a woman within this context. While I think there is a widespread assumption that gender-influences on academic careers have been largely resolved, in my experience this assumption is not justified. I argue that the reluctance to acknowledge the influence of gender on individual experiences and career success restricts what we can achieve in western academic environments, and, furthermore, limits insights into our strengths and weaknesses as women scientists. Finding the resonance between who we are and what we do was described by Palmer (1998) in *The Courage to Teach* as "...subtle dimensions of the complex, demanding, and lifelong process of self-discovery" (p. 13). Part of my self-discovery involved understanding and appreciating my strengths as a woman.

My process of self-discovery began in the dynamic social context of the late 20th century. I am part of the first generation of women who had it all. As a child of the baby boom generation, I benefited from emerging changes and opportunities for women and girls brought about at that time. Growing up in the midst of that social change impacted my development and self-concept as a woman. As a young woman I internalized and rebelled against lingering limitations and powerlessness of women in our society and never fully connected with the feminine. This is important to my story because so much of my development and emergence as a leader have been influenced by discovering the power of women's connectedness.

During my late teens and early twenties I worked to create an image of independence and toughness. I went far away to college, ran marathons, lifted weights, and took up rock climbing and hunting, not entirely unusual pursuits for a young woman at that time, but not entirely mainstream either. During and immediately after college I had several jobs as a research assistant working in remote backcountry locations on bird surveys and vegetation analyses. I completed my undergraduate

R. Bingham (✉)
Department of Natural and Environmental Sciences, Western State College, Gunnison, CO, USA
e-mail: rbingham@western.edu

K.C. Wieseman, M.H. Weinburgh (eds.), *Women's Experiences in Leadership in K-16 Science Education Communities, Becoming and Being*, DOI 10.1007/978-90-481-2239-4_5,

degree in botany and after working for several years in the field of environmental education I went back to graduate school and completed my master's degree in education. Although I was very interested in biology and had had some experience with research, my plan was to teach high school science because I enjoyed interacting with people and felt that the isolation and tedium of research would not be emotionally or socially satisfying over the long term. However, I loved biology and as I entered my late twenties, I could see the window of opportunity for doctoral study narrowing. The prospect of entering a Ph.D. program was daunting, but intoxicating. Once I made this decision, there was never a doubt that it was the right choice. I had found my passion and I was hooked.

Given my past experiences as a research assistant, I was comfortable with the techniques and approaches used in field ecology. However, my laboratory experience was much more limited, and as a result, I was insecure and extremely self-conscious working in that environment. The first lab in which I worked was dominated by male graduate students with a male faculty advisor. There was a lot of bantering and teasing. Any laboratory blunder was an opportunity to poke fun. It was all good-natured, but you had to be able to take it and you had to be able to dish it out. For me it was extremely stressful. It was a rough introduction to my doctoral program. I decided to change faculty advisors and labs. It was in this new lab, consisting entirely of female graduate students, that I discovered a woman's world. Although our advisor was male, his leadership and advising styles were very different from my previous advisor. The lab was clean and organized with a completely different culture than I had experienced before. I did not feel like I had to pretend that I knew how to do things; it felt okay to ask for help, and it felt okay to make mistakes. I could actually talk to my lab mates without posturing, joking, and ribbing. We still had lots of fun, but it was a different kind of fun. It was the fun of interacting closely and openly, creating relationships built on sharing stories, trusting, and opening up. It was about dropping pretenses and was a very nurturing environment within which I was relaxed enough to discover that I really enjoyed research.

This discovery however, did not override my primary identity as an educator. All through my doctoral program, I envisioned myself heading to a primarily teaching college. I had never seriously considered following a hard-core research track for my career. My identity as an educator guided me as I applied for jobs and helped me through the interview process itself. As I applied for jobs at teaching colleges, I viewed my academic background in education as a strength. However, it was during one of those interviews that I became aware that my master's degree in education could be viewed as a liability rather than an asset. Not having a master's degree in biology suggested to some individuals that I was not serious about real science. My path was untraditional and seen as a red flag. I was actually told as much in an interview. That may have been the first indication of the culture into which I was entering – the world of a science department in higher education in which the values and language of education are ignored at best and openly ridiculed at worst. Despite this situation, my background and commitment to education has, over the long run, served me well. As the world of higher education changes, I have been able to tap into it as a resource for understanding and embracing these changes, whereas faculty with a strictly sciences background find these changes difficult to understand and accept. It has not been easy, however, and my unique academic background has often contributed to feelings of marginalization.

Perhaps my greatest frustration working within higher education has been struggling within our current academic model (the transmission model of education) to create a truly collaborative learning community. Palmer (1998) described a community in this way, "Community is an outward and visible sign of an inward and invisible grace, the flowing of personal identity and integrity into the world of relationships" (p. 90). Unfortunately, if the goal is growing a community, as defined above, the structure of our system of higher education is fundamentally flawed. As noted by Nelson (2009) in her chapter in this book, community is a rare state among higher education faculty. Our system of higher education has grown from an objectivist world view, a philosophical framework that isolates rather than connects. As a young assistant professor in a department of science I did not understand this. Unaware of the world I was entering, I enthusiastically and eagerly invested myself into the work of creating a community of scholars assuming that my colleagues had the same goal. There was verbal affirmation of that goal by everyone in the department, so it was not unreasonable that I acted on it.

Creating a community involves bringing people together. Everything that I did professionally, beyond the mechanics of teaching, professional development, and service (in other words, beyond the expectations for tenure), grew out of this commitment. I organized and participated in a variety of activities for which the central goal was creating community. These activities included a science seminar series, potlucks in the dorms, freshman field trips, department picnics, and Biology Club. These efforts took an enormous amount of energy. Initially, my primary collaborators on many of these projects were two other junior faculty, also women. It did not escape our notice that the energy behind these efforts was largely, if not exclusively, our own.

From one perspective perhaps this was not an issue. I took on these projects because they were important to me, perhaps more of a priority for me than for my male colleagues. I chose to put my energy into these efforts. Unfortunately, rather than creating a community, these efforts left me feeling isolated and exploited. It began to seem that, rather than working on these projects with my colleagues, I was doing this work for my colleagues. At least that is how it felt. Everyone agreed that the projects and initiatives were valuable, but the effort to keep them going largely fell in my lap, and, while some participated regularly, others did so only when it was convenient, or not at all. After a few years it began to feel like I had unintentionally stepped into the traditional female role of serving others. While I am sure my efforts to build a community were appreciated, it is clear to me now that my efforts to build community ultimately undermined my continued professional development as a scientist, since these projects, along with a heavy teaching load and committee responsibilities, left little time and energy for research.

This time and choices were a critical step in my career where mentoring could have made a big difference, and where I believe gender played a role. A mentor could have helped me focus my energy and could have helped me to see where, within the context of my particular department, my particular institution, and given my professional goals, my energy would have been best spent. Perhaps a mentor could have helped me strategize ways to facilitate fuller participation by my colleagues. Moreover, a mentor who was also a woman may have better understood my impulses and, as a result, been able to communicate with me more effectively about balancing my time and energy than a man. Why did the senior faculty not perceive what was happening and intervene? Because I am a woman did my colleagues assume that other aspects of my career like research were less important to me than to others? Was this an example of unconscious sexism in that, because I am a woman it was somehow my responsibility to carry the burden of these projects and that it was okay for others not to follow through with responsibilities or politely excuse themselves from helping or participating? There is no way to know whether a female mentor specifically would in fact have made a difference. Regardless, as is often the case for junior faculty who are women, I navigated the years before tenure without a mentor and this had a negative impact on my career as a scientist. Lack of mentoring such as I experienced is not uncommon for women in academia and contributes to the challenges women face in this professional environment (Etzkowitz & Gupta, 2006).

As I reflect on this experience, I think that what may have contributed to the problem is not that my male colleagues do not value community. Rather what we experience and identify as community is fundamentally different and, moreover, our expectations for community in the workplace are also very different. These differences help to create the chilly climate of science (Barres, 2006; Bird, Litt, & Wang, 2004; Callister, 2006; Collins, 1998; Martinez et al., 2007; Murphy, Steel, & Gross, 2007; Settles, Cortina, Malley, & Stewart, 2006). This was a tough lesson to learn. It led to a few years during which I struggled with anger, depression, and bitterness. I felt exploited, taken advantage of, marginalized, and even ridiculed. I felt that I had made a mistake and that I did not belong in this career into which I had personally invested so much. I was seriously considering leaving higher education, when over the holidays one year not long ago I picked up Palmer's *The Courage to Teach*. My connection with the experiences Palmer described in the book was visceral. Reading that book helped me recognize that science and teaching are central to who I am and all I have to offer the world professionally as a scientist and as a woman. I could not leave, and I realized that I had to step up and be a change agent for institutional transformation if I were to stay. I explore some of my initiatives toward institutional transformation in the section on leadership activities at the end of this chapter.

Theoretical Framework

Authority and Underrepresentation of Women in the Sciences

Challenges facing women in science have been explored extensively over the past twenty years and a variety of programs have been developed to address the issue of the leaky pipeline (Bystydzienski & Bird, 2006; Cronin & Rogers, 1999;

Rosser, 2004). At some levels these efforts have been successful with women now making up significant proportions of the doctoral degrees awarded in science, mathematics, and engineering (Long, 2001; Xie & Shauman, 2003). However, despite these important gains, it is clear that challenges remain as women continue to be underrepresented in the ranks of tenured professors and academic leadership positions in colleges and universities (Long, 2001; Xie & Shauman, 2003). A variety of explanations have been proposed for why women either do not pursue academic careers in science or leave careers once they have begun. These explanations include the lack of mentoring, the challenges of dual-career couples and the demands of balancing work and family. Another challenge that has been explored within the science education and feminist literature, but has received less attention from the broader community, is the challenge of working within the prevailing and persistent masculine culture of science. In the paragraphs that follow I explore my experiences as a woman in the sciences and science education in the context of the research about underrepresentation of women in the sciences.

My experience actually provides an interesting perspective on this problem. I do not have the challenge of accommodating the career goals of my spouse. My husband's work situation is flexible and therefore I have been fortunate that he can accommodate the demands of my career. This has also been significant in the context of child care as he has been able to take over many of those demands as well. To some extent the classic sex roles are reversed in my family, which has relieved me of some of the pressures that are often given as reasons for women leaving the sciences. However, it has still been a difficult path for me as a woman. Recent insights from my teaching contribute to my understanding of this difficulty.

For years I have focused on developing a very student-centered approach to teaching, giving college students the lead in the classroom, while I take the role of facilitator. I prefer to teach this way, partly because I am not entirely comfortable taking center stage, but also because there is evidence that student achievement can improve under these circumstances (Prince, 2004). As an educator I enjoy the creative challenge of developing engaging classroom activities, and I appreciate the break from giving lectures. As a student myself, I always preferred doing things to just reading about them or hearing about them in a lecture. Doing enhanced my learning, and despite the fact that it required more effort, was much more engaging and rewarding. However, in my experience, and based on feedback from course evaluations, students do not necessarily see it this way. Students, some more than others, want authority. They perceive a teacher-centered approach (directive and authoritative) to course structure as better teaching. This may be particularly true of science students who have excelled within traditional classroom settings. It also may be particularly true for some demographic groups more than others. For example, men tend to expect and respond well to authority and may interpret efforts to downplay the hierarchical structure of a traditional college classroom as evidence of an incompetent instructor (Basow, 1995; Tannen, 1994). In my experience, some young men find in this setting an invitation to challenge or undermine the instructor's authority. While women are more likely to appreciate a more egalitarian and collaborative classroom style, this approach may still challenge their expectations

(Tannen, 1994). They may perceive the non-authoritative approach as nice but also less competent (Basow, 1995). At my institution, as at most teaching colleges, student evaluations of teaching are critical to promotion and tenure. Over the years, my evaluations have fluctuated substantially. I interpret some of this variation to my non-traditional teaching style. My experience with both students and colleagues is that a collaborative and experiential approach to learning goes against the prevailing authoritative classroom culture that persists, despite significant efforts for reform, in higher education, especially in the sciences.

Some of this variation in my teaching evaluations over the years may also be attributable to the fact that I am a woman. Though often dismissed, there is strong evidence for gender bias in student evaluations of professors (Andersen & Miller, 1997; Basow, 1995; Basow, 2000; Massoni, 2004; Miller & Chamberlin, 2000). This may be particularly true in fields that are traditionally male dominated, and among students, especially men, who come to college with traditional or conservative attitudes toward gender roles (Basow, 1995). In contrast to the national trend of women constituting the majority of undergraduates students at many colleges and universities, at my institution more than half of our students are men. In biology, approximately half of our incoming students hope to work in the field of wildlife management. Most of these students are male. Some of our students are the first in their families to attend college. Students from such backgrounds may hold expectations for gender roles that are more traditional than average. Over the years, some of my most painful teaching experiences have arisen in this setting. While we often hear about challenges to women from male colleagues and the faculty culture, I wonder to what extent challenges from students contribute to the chilly climate female faculty experience in higher education. Moreover, considering the weight given to student course evaluations in the context of tenure decisions at many institutions this is not a trivial concern as it may indicate a consistent source of gender bias in promotional criteria.

For me, further insights into these challenges have come from trying to unravel one of my most persistent problems – clearly communicating my expectations to both students and colleagues. The introspection and research involved in writing this chapter have helped me see that my conversational habits may have the effects of deemphasizing my authority and abrogating my actions as a leader within the current academic culture, leaving me feeling unheard, unappreciated, and impotent (Tannen, 1994). For example, my natural response to students (and colleagues) with problems is to try to help them feel better. When their discomfort is the result of an experience in my course, or if I am making a request of a colleague (rather than just doing the work myself), I feel the need to apologize. How interesting to discover that this need has been found to be a consistent conversational style among women (Tannen, 1994). However, apologizing has three insidious effects. First, it removes the responsibility for dealing with the problem from the student and where that responsibility lies becomes vague and ambiguous. Second, it diminishes my authority by equalizing rank across the interaction. Third, as a consequence of the first two effects, my expectations are less clear. Tannen (1994) described this tendency to respond empathetically in these types of social interactions as a general social convention among women, and also elaborated how these conversational conventions can be misinterpreted in some social contexts. The empathetic response equalizes rank across a social interaction, builds emotional connection, and is characteristic of women's interactions cross culturally (Fisher, 1999). Unfortunately the empathetic response can backfire in social contexts where the expectation is for authority.

At a meeting of our institution's women's faculty group, the education faculty had a name for this empathetic response. They called it the inner apology, in contrast to the inner authority – and we all nodded and murmured in recognition. When in a leadership position, this conversational habit can miscommunicate expectations. No wonder I cannot get anyone to live up to my expectations! My female colleagues helped me see that if you teach or lead from your inner apology rather than your inner authority, you risk losing credibility in your students' and colleagues' eyes. This may be the single most helpful and insightful advice I have received in nearly two decades of college teaching!

Genderization and Objectivism

In trying to understand my experiences in and out of the college classroom I began to explore some of the classic feminist literature on gender and society. This exploration also emerged from my experience as a parent and my closeness to the evident differences between boys and girls that emerge early in development. These developmental differences have been widely documented in the literature (Cahill, 2006; Fisher, 1999; Goldman-Rakic, 1995; Hines, 2004; Sax, 2006). According to Fisher (1999), boys typically develop very hierarchical social structures that emphasize competition in contrast to the more egalitarian social structures used by girls that value sensitivity to one another's needs and feelings. These differences persist in the conversational conventions and rituals of adults (Tannen, 1994). As I experienced in graduate school and now often experience at work, men interact with jokes, teasing, and playful put-downs. In contrast, women's conversational rituals take into consideration the "...effect of the interaction on the other person, and ... downplay the speakers' authority so they can get the job done without flexing their muscles in an obvious way" (Tannen, 1994, p. 23). These words resonate with my experience. While both conversational styles are effective, the social setting within which these interactions occur may influence how an individual is perceived.

Within the traditionally male dominated environment of science, a woman's conversational strategies that are designed to "... cultivate closeness, connection, consensus and confirmation" (Fisher, 1999, p. 35) have the effect of making her appear less competent than she really is and can impact her effectiveness as a leader. Most academic environments, but science in particular, is based on a culture that values objectivity, the way of knowing that defines truth as known only through our disconnection with that which we wish to know (Keller, 1985). In *Reflections on Gender and Science* Keller (1985) described the emergence of objectivity as the dominant mode of scientific thought during the 17th century. The institutionalization of modern science at this time involved the domination of the mechanical philosophy which sought to separate mind from spirit, and took place against a backdrop of dramatic social and cultural upheaval. Keller argued eloquently that prevailing concepts of gender also shifted at this time, such that concepts of male and female became much more highly polarized than they were previously. She argued that the domination of the mechanical philosophy succeeded in cementing the polarization between masculine and feminine "... that was central to the formation of early industrial capitalist society" (p. 20) and still prevails in the genderization of science today, and the "... historically pervasive association between masculine and objective, more specifically between masculine and scientific" (p. 75).

Palmer (1998) described objectivism as our fearful way of knowing, and saw it as the prevailing paradigm of teaching and learning in our culture and also our primary obstacle to teaching and learning. Echoing the ideas of Keller, he proposed that objectivism disconnects us from our subjects, our students, and our colleagues, and ultimately prevents us from discovering the truths we seek as scholars. Furthermore, both Keller and Palmer made a powerful argument for the dependence of knowledge, in its deepest sense, on relationship and connection rather than on distance and separation.

Relational, connective, empathetic, intuitive, contextual, emotional, and subjective, all have been words used to describe (at times to denigrate) feminine qualities and to explain why a woman's brain is inherently less suited for science as compared to a man's brain. Interestingly, these words also were used to describe the untouchables (individuals whose jobs could not be outsourced) in Friedman's *The World is Flat* (2005). Similarly, Daniel Pink, author of *The Whole New Mind* (2005), predicted that individuals exhibiting these qualities will rule the 21st century. In describing the qualities and character traits necessary for success in our emerging conceptual world, Pink referred to the left brain/right brain dichotomy of popular psychology and argued that individuals with more right brain attributes such as intuition and people skills will be most successful in the coming century. Although greatly oversimplified in the popular literature and used metaphorically by Pink, there is ample evidence that some mental processes do occur primarily on one side of the brain or the other, and that these differences are influenced by gender (Hines, 2004; Kimura, 1996; Witelson, 1991).

Recent findings from the fields of neuroscience and genetics provide further support for the reality of biological differences in brain function between men and women, including brain differences affecting traits such as memory, vision, hearing, interpretation of facial expressions, and emotion (Cahill, 2006; Goldman-Rakic, 1995; Holyoak & Kroger, 1995; Skuse et al., 1997). That these different capacities may have a heritable component does not surprise me as an evolutionary biologist, nor threaten me as a feminist. Humans, like other animals have important reproductive asymmetries between males and females. Through our evolutionary history, as a result of these reproductive asymmetries, males and females have experienced different pressures through natural selection, which in turn have given rise to innate differences between the sexes (Morelli, 1997).

As both biological and social phenomena, it is interesting to contemplate the implications of these gender differences for the future of science and for the future of women. Keller (1985) explored these possibilities through her fascinating examination of the life and work of Barbara McClintock. McClintock was famous for her discovery of transposable elements. She won the Nobel Prize in 1983, but was marginalized in the world of science throughout most of her career. Her marginalization arose not just due to the fact that she was a woman, but because her approach to science was considered eccentric and unorthodox. Relational, connective, empathetic, intuitive, contextual, emotional, and subjective – these terms, referred to above, were also used to describe McClintock and her work. Keller argued that in McClintock we can find an approach to science not based on division, separation, and pure objectivity, but one based on connection, relationship, and empathy. McClintock provided a vision of how women may do science as women, or more profoundly, do gender-free science. In this view the focus is not on creating sameness among scientists, but on the value brought to science by diversity of practice. This, of course, is the problem with most of the interventions that to this date have been proposed and implemented to enhance opportunities for women in science and to retain women in science careers. The focus has always been on fixing women so they will fit within the strongly masculine culture that characterizes scientific and technological workplaces (Henwood, 1996; Wajcman, 1991). These same ideas were proposed by French philosopher and psychoanalytic theorist Luce Irigaray who sought to develop a concept of woman that is not defined in terms of relationship to man, but that is distinct and unique on its own terms (Gilbert, 2001). This concept of woman as distinct on its own terms speaks loudly to me and I can begin to envision my strengths as a woman operating in this emerging world of science and science education.

Leadership Activity and Response

My emergence as an academic leader has grown from the challenges I have experienced as a woman and a desire to see change. There have been a few major challenges that have drawn me out as a female leader and, in fact, left me no choice but to step into a leadership role. I share them in the form of a personal story so you can consider their significance and relate them to your local context. Some of the descriptions and stories for the challenges are more developed with concrete details for their implementation, while others are broad conceptualizations. The extent of explanation associated with each leadership challenge parallels my being and becoming leadership journey. As you will read, in my experience the key challenges to academic leadership by women are

(1) The controversial topic and use of student course evaluations in hiring and academic career advancement
(2) The central responsibility of senior faculty in both formal and informal leadership positions

(3) The reflective journey of self-awareness of one's strengths and challenges as a leader

Challenge 1, Student Course Evaluations

The first challenge centers on the controversial topic of student course evaluations and is tightly linked to Challenge 2, that of effective mentoring. Issues related to the role of an academic leader in the context of student evaluations, which I have faced, include responding to concerns such as:

- How does a young assistant professor maintain integrity as an educator in the face of the evaluations?
- How does inexperienced faculty avoid caving into the temptation to keep the students happy at the expense of academic rigor?
- How does senior faculty deal with new faculty falling into the trap of feeling like they have to be nice in an effort to raise evaluation scores?
- How does faculty help instructors deal emotionally with the not infrequent ugly comments?
- Are my concerns about gender bias on course evaluation scores real?

I came to my current institution directly after finishing my Ph.D. I remember coming to my new department brimming with confidence and optimism; I was fearless. Unfortunately, after a very short time I became overwhelmed with feelings of inadequacy and consumed with the fear of failure. Some of this was a result of my own insecurities and poor or inadequate mentoring, but I think some can also be attributed to the culture of fear (Palmer, 1998) that pervades higher education; a culture that separates rather than connects. My personal fears dealt largely with my performance in the classroom, fearfulness of my students, concerns that I was not liked, and that I was being too tough on them or not tough enough. I feared talking to my senior colleagues about these challenges and was not sure about how to work within the system to get the help I needed at that stage of my career. It had been made quite clear to me when I was hired that student evaluations of teaching were of paramount importance in hiring and tenure decisions, and so I agonized over them each semester. The more I agonized, the more anxious, self-conscious, and awkward my teaching became. I have come to terms with this sometimes painful aspect of my job, and I think this facet of academic careers at teaching colleges is one for which insightful and proactive mentoring is critical. While I do not think student evaluations need to be completely eliminated, I do think that there is need for major changes in the system. I also think that women and men may respond differently when confronted with student evaluations at the end of a term and may need different support systems as they work through what can be an emotionally draining experience. Furthermore, if these evaluations contribute to hiring, tenure, and promotional criteria, we need to be aware that the system may discriminate against female instructors because of a cultural bias in the classroom.

Within my department I have taken on the leadership activity of initiating discussions about the problems associated with student course evaluations and their potential biases. This has been one of the most uncomfortable experiences in my career and I have been surprised at the intensity of argument from both proponents and critics. Despite this difficulty, persistent questioning of the use of our evaluation instrument from voices across campus has resulted in recent important modifications in how it is used for faculty evaluation. There may be opportunity at your institution to initiate discussions to address concerns about student course evaluations. These discussions might lead to changing the evaluation instrument itself or the role the instrument plays in decisions related to hiring, tenure, and promotion. For such an initiative it can be important to find allies within your department or institution with whom to collaborate, since, in my experience discussion of and change related to student course evaluation is a topic that many academics would rather avoid. Critical mass can help get the information out to a broader audience for the consideration it is due.

Challenge 2, Responsibility of Senior Faculty

As I move into the ranks of senior faculty, I am confronting the challenge of effective mentoring. Our institution is poised to develop a more thoughtful and comprehensive mentoring system, a project with which I look forward to working. In the meantime as an individual, I try to implement good mentoring practice with my junior colleagues by sharing syllabi and teaching strategies, including them in faculty discussion groups, meeting over coffee to commiserate and strategize over student evaluations, and generally being available when they have questions or need support. More importantly, I see it as my responsibility to protect junior faculty from overextending themselves in the same potentially unproductive paths that I wandered when I was in their shoes. With three new members of the biology faculty that are women, I try to provide the kind of support and direction that I feel I lacked. These actions as an individual faculty member comprise a leadership activity which you can consider and unfold at your institution.

As I move into the ranks of senior faculty, I have been faced with different leadership opportunities. Opportunities for leadership come in many contexts. Leaders are not just those formally designated within the administrative structure of the institution. Some of the most powerful leaders are those that lead from within, those that grasp a vision and toward which they rally their colleagues. Of course, if one desires real change, taking advantage of formal leadership opportunities is critical. In my situation, I volunteered to take on the responsibility of biology discipline coordinator within my department. At our institution this position oversees program structure and function including such things as scheduling, hiring, and budget. Unlike a department chair, however, the role and position come with no compensation either in the form of extra salary or reassigned time. As a result, no one is particularly eager to add it to an already overloaded schedule. I could see that this formal leadership opportunity would provide the avenue to implement the organizational changes that I felt were important in moving our program in a progressive direction; a direction with greater awareness and sensitivity to issues women face within the academic environment, a direction promoting effective mentoring, a direction more supportive of innovative pedagogy, and a direction friendlier to thoughtful program assessment.

As the biology discipline coordinator I faced several challenges, including recruitment and retention of high quality faculty, development of an effective faculty mentoring program, scheduling and teaching load conflicts, development of quality program assessment tools, and the ubiquitous challenge of offering a high quality curriculum with limited resources. There were also concerns regarding shared ownership of the co-curricular aspects of the program which seemed to regularly fall into the laps of one or two faculty, but which would have greatly benefited from wider participation by all. Organization and, more importantly, coordination of the different aspects of our program were needed, but our program faculty seemed to lack the collective will or discipline to accomplish anything. It was difficult to convince colleagues of the need to meet regularly to allow for the thoughtful discussion that was necessary to make appropriate changes. There was an atmosphere of complacency which left me overwhelmingly frustrated. The program culture within which I found myself was not particularly open to critical self-reflection and change. However, I can point to several significant accomplishments. We made progress streamlining our schedule to more efficiently utilize space and deliver our laboratory curriculum. We hired our first laboratory coordinator, and we successfully filled one tenure-track position. As the discipline coordinator I organized, with the support of my department chair, a faculty retreat that provided the opportunity for concentrated discussion of many of the challenges described above. While we have yet to implement some of these changes, we were successful in taking the first steps in what I feel is a positive direction.

Emerging as a senior faculty leader under these conditions was difficult and my year in that position was pretty rocky. I have had to confront and examine my weaknesses as a leader with the hope that these insights will allow me to grow into a positive leadership role. Effective leadership in any context requires disciplined reflection and self-awareness, the third challenge to women in academic leadership positions.

Challenge 3, Reflection about Own Strengths and Challenges as a Leader

Personal reflection has been essential as I have confronted and struggled to move beyond my limitations as an academic leader, and as I have learned to recognize and celebrate my strengths as a leader. Though personal reflection is critical to our development as leaders, it can be hard to find the intellectual space for this mental activity in our busy lives. I was fortunate to have the opportunity to participate in Project Kaleidoscope (PKAL), an NSF-funded initiative supporting institutional transformation in the STEM fields. Being involved in PKAL has provided me with support, inspiration, and opportunities for personal reflection. I encourage you to be intentional about your development as a leader and to seek out opportunities like those offered through PKAL. It is unfortunate that many of us have the idea that leaders are born rather than made, since this leaves so much of our leadership potential untapped. Taking advantage of opportunities for leadership skills development can enhance and accelerate your ability to affect change.

While it is important to reflect on both your strengths and challenges as a leader, I recommend that you look at your strengths first and celebrate them. Keep in mind that your strengths may not measure up on the current scale of scientific achievement. As women, we are finding voices that are not defined in terms of relationship to men, but are distinctly our own. To this end, networking with other women faculty is valuable for sharing experiences and perspectives that may be unique. Across the country many colleges and universities have established women faculty groups for this very purpose. The women's group on my campus is in its infancy. I initiated this activity with my female colleagues across campus as an informal discussion group that meets approximately once a month. Perhaps this too is a leadership activity that could be implemented at your institution.

Despite my frustrations and setbacks as a woman in science, I am optimistic as I look ahead. The world of higher education in western academic environments is changing as numerous recent initiatives promoting institutional transformation and reform in undergraduate science education indicate. Reassuringly, the academic culture is becoming increasingly open to self examination. Some of this change is the result of a younger generation swelling the ranks of faculty; some is the response to persistent voices calling for change. I can see how my particular combination of skills, training, and interest fit into this emerging world. I feel confident that I have much to contribute. While challenges remain, I feel fortunate to be a scientist, an educator, and a woman at this point in history when there are so many opportunities for us, as women, to lead.

References

Andersen, K., & Miller, E. D. (1997). Gender and student evaluations of teaching. *Political Science & Politics, 30*, 216–219.

Barres, B. (2006). Does gender matter? *Nature, 442*, 133–136.

Basow, S. A. (1995). Student evaluations of college professors: When gender matters. *Journal of Educational Psychology, 87*(4), 656–665.

Basow, S. A. (2000). Gender dynamics in the classroom. In J. C. Chrisler, C. Golden, & P. D. Rozee, *Lectures on the psychology of women* (2nd ed., pp. 44–55). New York: McGraw-Hill.

Bird, S., Litt, J., & Wang, Y. (2004). Creating status of women reports: Institutional housekeeping as "women's work." *NWSA Journal, 16*(1), 194–206.

Bystydzienski, J. M., & Bird, S. R. (2006). *Removing barriers: Women in academic science, technology, and mathematics (STEM) careers*. Bloomington, IN: Indiana University Press.

Cahill, L. (2006). Why sex matters for neuroscience. *Nature Reviews Neuroscience, 7*(6), 477–484.

Callister, R. R. (2006). The impact of gender and department climate on job satisfaction and intentions to quit for faculty in science and engineering fields. *Journal of Technology Transfer, 31*, 367–375.

Collins, L. H. (1998). Competition and contact: The dynamics behind resistance to affirmative action in academe. In L. H. Collins, J. C.Chrisler, et al. (Eds.), *Career strategies for women in academe: Arming Athena* (pp. 45–79). Thousand Oaks, CA: Sage Publications, Inc.

Cronin, C., & Rogers, A. (1999). Theorizing progress: Women in science, engineering, and technology in higher education. *Journal of Research in Science Teaching, 26*(6), 637–661.

Etzkowitz, H., & Gupta, N. (2006). Women in science: A fair shake? *Minerva, 44*, 185–199.

Fisher, H. (1999). *The first sex: The natural talents of women and how they are changing the world.* New York: Ballantine Books.

Friedman, T. L. (2005). *The world is flat: A brief history of the twenty-first century*. New York: Farrar, Straus and Giroux.

Gilbert, J. (2001). Science and its 'other': Looking underneath 'woman' and 'science' for new directions in research on gender and science education. *Gender and Education, 13*(3), 291–305.

Goldman-Rakic, P. S. (1995). Architecture of the prefrontal cortex and the central executive. In J. Grafman, K. J. Holyoak, & F. Boller (Eds.), *Structure and functions of the human prefrontal cortex. Annals of the New York Academy of Sciences 769* (pp. 71–83). New York: New York Academy of Sciences.

Henwood, F. (1996). WISE choices: Understanding occupational decision-making in a climate of equal opportunities for women in science and technology. *Gender and Education, 8*(2), 199–214.

Hines, M. (2004). *Brain gender*. New York: Oxford University Press.

Holyoak, K. J., & Kroger, J. K. (1995). Forms of reasoning: Insight into prefrontal functions. In J. Grafman, K. J. Holyoak, & F. Boller (Eds.), *Structure and functions of the human prefrontal cortex. Annals of the New York Academy of Sciences 769* (pp. 253–263). New York: New York Academy of Sciences.

Keller, E. (1985). *Reflections on gender and science*. New Haven, CN: Yale University Press.

Kimura, D. (1996). Sex, sexual orientation and sex hormones influence human cognitive function. *Current Opinion in Neurobiology, 6*(2), 259–263.

Long, J. S. (Ed.) (2001). *From scarcity to visibility: Gender differences in the careers of doctoral scientists and engineers*. Washington, DC: National Academy Press.

Martinez, E. D., Botos, J., Dohoney, K. M., Geiman, T. M., Kolla, S. S., Olivera, A., et al. (2007). Falling off the academic bandwagon: Women are more likely to quit at the postdoc to principal investigator transition. *EMBO Reports, 8*, 977–981.

Massoni, K. (2004). The influence of gender on students' evaluations of teachers or why what we can't count can hurt us. *Sociologists for women in society*. Retrieved February 15, 2008, from http://www.socwomen.org/resources/genteachevalfactsheet.pdf

Miller, J., & Chamberlin, M. (2000). Women are teachers, men are professors: A study of student perceptions. *Teaching Sociology, 28*, 283–299.

Morelli, G. (1997). Growing up female in a farmer community and a forager community. In M. E. Morbeck, A. Galloway, & A. L. Zihlman (Eds.), *The evolving female: A life history perspective* (pp. 209–219). Princeton, NJ: Princeton University Press.

Murphy, M. C., Steel, C. M., & Gross, J. J. (2007). Signaling threat: How situational cues affect women in math, science, and engineering settings, *Psychological Science, 13*, 879–885.

Nelson, T. (2009). Building leadership capacity by nurturing community. In K. C. Wieseman & M. Weinburgh (Eds.), *Women's experiences in leadership in K-16 science education communities: Becoming and being* (pp. 167–181). Dordrecht, The Netherlands: Springer Publishing.

Palmer, P. J. (1998). *The courage to teach: Exploring the inner landscape of a teacher's life*. San Francisco: Jossey-Bass.

Pink, D. H. (2005). *A whole new mind: Why right-brainers will rule the future*. New York: Riverhead Books.

Prince, M. (2004). Does active learning work? A review of the research. *Journal of Engineering Education, 9*(3), 223–232.

Rosser, S. (2004). *The science glass ceiling: Academic women scientists and the struggle to succeed*. New York: Routledge.

Sax, L. (2006). What teachers need to know about the emerging science of sex differences. *Educational Horizons, 84*, 190–212.

Settles, I. H., Cortina, L. M., Malley, J., & Stewart, A. J. (2006). The climate for women in academic science: The good, the bad, and the changeable. *Psychology of Women Quarterly, 30*, 47–58.

Skuse, D. H., James, R. S., Bishop, D. V. M., Coppin, B., Dalton, P., Aamodt-Leeper, G., et al. (1997). Evidence from Turner's syndrome of an imprinted X-linked locus affecting cognitive function. *Nature, 387*, 705–708.

Tannen, D. (1994). *Talking from 9 to 5: How women's and men's conversational styles affect who gets heard, who gets credit, and what gets done at work*. New York: William Morrow and Company, Inc.

Wajcman, J. (1991). *Feminism confronts technology*. London: Polity Press.

Witelson, S. F. (1991). Neural sexual mosaicism: Sexual differentiation of the human temporoparietal region for functional asymmetry. *Psychoneuroendocrinology, 16*, 131–153.

Xie, Y., & Shauman, K. A. (2003). *Women in science: Career processes and outcomes*. Cambridge, MA: Harvard University Press.

Part III
Transplanting and Growing: Physical-Geographic Relocation and Accompanying Huge Cultural Shifts

Chapter 6
Resolving Issues of Cultural Discontinuity in Co-facilitation of Professional Development Activities

Hedy Moscovici and Irene Osisioma

Stories of Experience

Setting the Stage

In 2004 we became actively involved in the planning and implementation of a series of professional development (PD) activities that focused on newly developed middle school science immersion units that span a 4–6 week teaching period. The participants were all inservice teachers working in the school district near our university. We served as the science education experts on the 8th grade and the 7th grade leadership/facilitation (L/F) teams respectively. The L/F teams included a science expert (SE) from the district and the local district (due to the large number of students, the district is divided into a number of local districts), a practicing lead science teacher (ST) who had implemented the unit successfully at the classroom level, a science faculty (SF) expert from the university, an expert in science PD from outside the area, and the science education faculty (SEdF). The goal of each L/F team was to provide inservice teachers with the best PD on how to use inquiry-based approach to science instruction in the middle school classrooms.

In this chapter, we tell our stories about our co-facilitator roles and participation in our respective L/F teams of this PD. Our intention is to provide readers with practical guides on how to grapple with the challenges of cultural conflicts and discontinuity that may arise during educational leadership activities such as professional development sessions. We focus on the relationships that affected the functioning and cohesiveness of our L/F teams as well as the challenges we encountered as members of these teams, specifically the interplay between cultural discontinuity and cultural conflict in the contextual milieu of PD.

We are both immigrants to the United States, although from different countries. Though we share the common experience of immigration, we have had dramatically different experiences due to our jobs, as well as exposure to various PD subgroups.

H. Moscovici (✉)
Division of Teacher Education, Center for Science Teacher Education, California State University, Fullerton, CA, USA
e-mail: hmoscovici@csudh.edu

K.C. Wieseman, M.H. Weinburgh (eds.), *Women's Experiences in Leadership in K-16 Science Education Communities, Becoming and Being*, DOI 10.1007/978-90-481-2239-4_6,

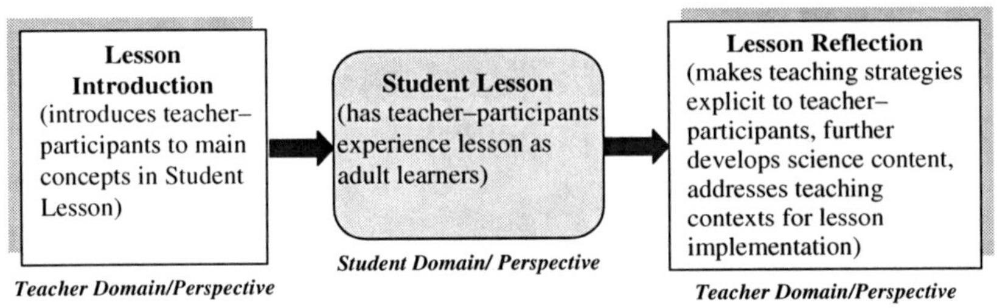

Fig. 6.1 Professional development lesson sequence

Hedy's story reveals little cultural conflict and discontinuity due to her long standing involvement in diverse capacities with the partner school district. Irene's story captures a different experience including conflict and resistance resulting from her unrecognized expertise as a new university faculty, limited involvement with the district leadership group, and her perspective as an insider (i.e., she taught in a partner high school). As a result our stories are very different despite the parallel PD experience of training teachers on how to implement the immersion units in their middle school classrooms. Hedy worked with the 8th grade team on an immersion unit on *Density and Buoyancy*; Irene worked with the 7th grade team using an immersion unit on *Variation and Natural Selection*.

The curriculum for the five-day immersion PD experienced by teacher participants was divided into steps that focused on a small number of science concepts. Each step consisted of a number of lessons presented in a pre-set lesson template sequence prepared by the PD organizers (see Fig. 6.1).

During the PD teacher participants experienced each lesson as both adult learners and teachers. All questions regarding classroom applications were discussed and answered in the Lesson Introduction and Lesson Reflection sessions. All questions pertinent to the lesson itself were addressed during the Student Lesson session. The PD experience itself provided the L/F teams with multiple opportunities to experience cultural conflict as this structure did not allow mixing the two teacher domains with the middle adult learner – student domain.

The analysis and evaluation of the PD by the external evaluator also provided the L/F teams with information revealing areas of success and needed improvements, as well as possible aspects of cultural conflict.

Hedy's Story

My first PD preparation session took place in the Center for Science Teacher Education on the university campus. In this session we planned for the upcoming five-day PD event and assigned roles and responsibilities to co-facilitators of the L/F team for the different PD segments. Most of the session participants had previously come to the center on different occasions such as the preparation stages for other co-lead PDs, and the development stages of the immersion units.

The first session was comfortable since we were beginning on common ground. The SEs from the local district were familiar faces. We had worked together for eight years on grant applications, PD implementations, and had attended common meetings geared toward the development of the immersion units. We began our planning session sharing past experiences in PD, reviewing data produced and shared by the external evaluator, and discussing the three-segment lesson template described at the beginning of this chapter.

During this session the conversation about planning the different facilitation segments of the actual PD experience flowed, and we all felt comfortable working together. Most PD lesson segments were assigned to a lead person who was assisted by another person, both referred to as co-facilitators. Roles for the implementation of the PD were rotated among co-facilitators and were flexible. We felt comfortable enough to say if we did not want to lead or participate in facilitating a certain segment of a PD event, because most of us (with the exception of the ST) had worked intensively (bi-weekly meetings) together over a year during the development stages of the immersion unit.

Below is an example of the kind of conversation that would take place during a typical leadership planning session. The conversation shows no signs of cultural discontinuity and the energy is focused on planning the PD event.

Dana (ST): So Hedy and I will take the introduction part and reflection parts of every lesson, while Rob (SF) and Alex (SE) will concentrate on the student lesson itself.

Hedy (SEdF): Fine with me. [looking around to include the other members] What do you think?

Rob (SF): No problem here. We are going to divide the roles so it will not be boring. I might take the intro to the student lesson and the reflection parts, while Alex will do the lesson with the participants.

Alex (SE): Should we keep the same role within the Student Lesson section?

Dana (ST): It is up to you, I think. [looking around] What do you think?

Hedy (SEdF): We should be flexible.

Rob (SF): Fine with me. I might be interested to have the part where I can pose challenging questions to reveal misconceptions; however, I am sure that Alex can do it too. So we can team up.

Dana (ST): Let's put some names [typing names and roles in the facilitator guide on her computer] and I will send a copy to you all. We'll talk after I have something more solid.

Hedy (SEdF): As long as you don't ask me to do something from one second to another, I am fine.

Dana (ST): I know, I know you like to plan and be prepared.

Hedy (SEdF) [looking toward Rob (SF):] At the university we like to have the time to prepare prior to class sessions...

Alex (SE): While at the district we need to be ready to deliver from one minute to another, depending on the dynamic waves of power...

During the PD, I had to lead some sessions at the last moment, while other planned sessions had to be modified (also at the last moment) according to the needs of the attending teachers, and various crisis moments during facilitation. I felt comfortable co-teaching and co-facilitating even with the last moment changes.

The wonderful atmosphere and collegial dialogue among our expert team as we planned the PD for middle school science teachers brought back memories of cultural discontinuity during my early years at California State University (CSU). I relate these early memories to portray two contrasting cultures. The former PD used a top-down and centrally controlled approach in leadership with the expectation of having university faculty direct PD for district teachers. The current PD was to be collegial, respectful, and facilitated rather than directed. When I began my career as a CSU faculty, a colleague (then the Chair of the Department) asked me to collaborate on a grant with the district and organize five-day PD experiences around science inquiry, English Language Learners, and science content standards. I submitted a grant application with a Co-Principal Investigator (Co-PI) from the district, and the grant was approved. Collaboration had a very different meaning then. I recall the grant Co-PI from the district coming for one day to observe PD activities. He sat in the back of the room for a number of hours, disappeared during the lunch break and never returned. As Principal Investigator of this PD grant, I took my role very seriously and organized a teaching team that would co-facilitate the PD together with me. I recruited according to expertise and availability from the university as well as from the district. The district personnel that I recruited were surprised by the collaboration as they were used to being told what to do. In the previous grant collaboration with the district there was a certain dose of cultural discontinuity between the university culture, where faculty were used to planning in advance and trying everything before the course session, and the school district culture, whose representatives were always ready to go and provide PD sometimes with only thirty minutes notice.

For this grant collaboration I designed common planning sessions with all participants to diminish or even alleviate cultural discontinuity and encourage cohesiveness among the members of the L/F teams. In addition, I scheduled individual meetings for university faculty to help them feel more comfortable with the subject that they needed to address in light of district standards and the needs of teachers, including inquiry science teaching and learning.

Irene's Story

I was raised in and completed all my formal education in Nigeria. In Nigeria the teacher/PD leadership members are considered experts and are not to be challenged, especially not openly. I arrived in the United States and started teaching science at the high school level with this mindset that I was stepping into an expert role and was prepared to provide all the information needed. It was not long before I experienced my first culture shock that students in the United States have as much power as their teachers, and sometimes more. The first thing I observed was that the concept of respect and authority was perceived differently from that with which I was familiar. In the Nigerian culture the teacher is viewed as the expert who has full authority and control of what goes on in her/his classroom. In contrast, in this new culture the teacher did not have full authority but had to negotiate how to run his/her classroom with school administration and students. My challenge was to assimilate into this new culture, and I needed to do so quickly to survive. My seven years of teaching science at the high school level therefore represented a struggle for survival. My greatest struggle was expressing my opinions during faculty meetings and other professional gatherings. I felt uncomfortable with the way discussions were held and so preferred to listen rather than take part. Before long, I discovered that my passivity was viewed as inadequacy or lack of expertise. I was forced to begin to speak out for myself and defend my expertise.

As a CSU faculty member, I was told that I would join a PD leadership/facilitation team as a science educator. I was very excited when I discovered that my role in this team would be to bring my expertise in science education to the table. I was excited to have the opportunity and a forum to share my experiences from 14 years of teaching science at the high school level (seven in Nigeria and seven in South Central Los Angeles) and my expertise as a science educator with the *Variation and Natural Selection* L/F team. I was going to be able to help the PD participants come to terms with inquiry-based instruction. A typical discussion that I had with the PD science teacher participants early in the project was:

Erin (ST): Irene, you said you were teaching science in a high school for some years. How do you think this [immersion unit implementation] will play out in the actual classroom?

Irene (SEdF): Pretty well. I am optimistic that this approach will help engage the students and we can begin to help them learn science the way they should.

Erin (ST): I am concerned that this approach is time consuming and we are expected to cover all the standards that will appear on the periodic assessment and California Standards Test (CST).

Irene (SEdF): You will not be expected to teach all the content, however if you use this approach to teach the immersion unit you should have equipped the students with the critical thinking skills to navigate the other content standards.

Erin (ST): Well, this too shall pass because we have had lot of strategies thrown at us in the past and now you are training us in another.

Irene (SEdF): It is unfortunate that you feel that way. I am optimistic, however, that if through inquiry-based instruction you develop students' critical thinking skills and teach them to enjoy learning science, for these students this will not pass. It will be something that they will enjoy and have forever.

Becoming a member of the *Variation and Natural Selection* L/F team in which every member held extensive expertise launched me into my next culture shock – a culture of power negotiation and reduction of cultural discontinuity. Working as a co-facilitator in this L/F team was not as rewarding as I expected. I got the feeling that my expertise was not valued by the way the team members ignored my contributions during the pre-PD preparations and during the pre-PD and post-PD daily sessions. This treatment continued until I had a chat with Scott, the SE from the district and the leader of our L/F team. Below is an excerpt from that chat:

Irene (SEdF): I just wanted to talk to you about how I have been treated by the team.

Scott (SE): Sure, I am very open to such discussions.

Irene (SEdF): I feel that my expertise is not valued. I see it by the way team members and especially you ignore my contributions during our pre-institute preparations and post-institute debriefs.

Scott (SE): I am so sorry that you felt that I ignored you in any way. I was not doing this on purpose and I am so sorry about it. Now that you have brought this to my attention, I will make sure to honor your contributions in the future. And indeed, I will become intentional about asking for your contributions.

Irene (SEdF): Thank you. I will appreciate this a lot because the reason I was brought into this project was to contribute my expertise as a science educator and a high school teacher. Not honoring my contributions makes me not to feel like part of the group and that makes me uncomfortable.

Scott (SE): I know! I promise to do things differently now that you have called this to my attention and I know how it is affecting you. Trust me; I am so sorry that I gave the impression that I ignored you in the first place. Thank you for letting me know how you feel and please feel free to call me to order when you feel that I begin to ignore your contributions in the future.

Theoretical Framework

The theoretical framework for this chapter finds its roots in three different areas of literature: (1) cultural discontinuity, (2) theory of human capital as applied to co-generative dialoguing and team-teaching, and (3) leadership development.

Cultural Discontinuity

We contend that cultural discontinuity as a concept applies to the work of professional development leadership teams, not just home school contexts. Our leadership work in professional development leads us to believe that the principles transfer when applied to negotiating personal culture and accepted protocols within professional development leadership teams, despite the fact that the concept has previously only been used to describe lack of cultural congruence and cohesiveness between home and schooling, whether the schooling is mandatory such as high school (Sheets, 1996), or elective such as university (Brunn, 1997). As immigrants to the United States we faced cultural discontinuity that was rooted in our cultural differences. Living in Romania and Israel prepared Hedy to speak up and be able to engage in evidence-based arguments in a professional, results-oriented manner. Working with the district for more than eight years prepared her to accommodate the tendency of teachers and district representatives on the L/F Team to change and renegotiate roles and responsibilities at the last minute. District representatives learned to provide data-based reasoning for last minute changes, thus reducing the extent of cultural discontinuity. Irene on the other hand, was a more recent immigrant and only knew a culture in which university professors were perceived as experts and not to be challenged. In her L/F team Irene struggled with finding a comfortable way to express her discomfort at not being valued for her expertise. "People are not only perceivers, knowers, and actors but are also self-reactors with a capacity for self-redirection" (Bandura, 1969, p. 677). As posited by Bandura, Irene underwent an introspective reflective process, which was brought to a level of consciousness without interrupting the already established group flow. She had discussion with Scott, the leader of her L/F team. Through this discussion Scott became aware of the ongoing discord within the group. As the leader of the team, he was able to re-establish needed group harmony and create a bridge to reduce cultural discontinuity. These actions brought about needed cohesiveness and effectiveness for PD co-facilitation.

We have applied the concept of cultural discontinuity (Phelan, Yu, & Davidson, 1994). In their work describing cultural differences between school and home, they noted a positive relationship between cultural discontinuity as related to students' performance and their self-esteem. The literature on cultural discontinuity in the United States covers a variety of groups such as American Indian (Brunn, 1997), African American (Diller, 1999; Davis, 1996), Asian Indian Americans (Beloo, 2002), and Latina and Latinos (Sheets, 1994). Sheets' (1996) research applied Gay's framework for interethnic group interactions – procedural or protocol-related [i.e., how of an interaction], substantive or content-related [i.e., what of an interaction], and interpersonal or people-related [i.e., who in an interaction]. This research revealed that the majority of high school classroom conflicts were of an interpersonal nature for African American and Chicano students, and a procedural and an interpersonal nature for Filipino American and European American students. Like these studies, the cultural discontinuity experienced in our PD project was also interpersonal and procedural in nature.

We suggest that cultural discontinuity (and possible conflict) may result from a combination of ethnic roots (Sheets, 1994) mixed with parents' and individual's participation in cultural traditions and worldviews (Brunn, 1997). Sheets (1994) showed that Latina and Latino high school students previously identified as at risk, performed at the level of gifted on the Advanced Placement Spanish Language Examination when provided with an instructional intervention focusing on positive teacher–student and student–student interactions and teaching–learning strategies that were culturally centered. This work suggests that focusing on actions that diminish cultural discontinuity has potential to decrease conflict and increase students' achievement. Brunn (1997) presented the story of two American Indian men who had grown up on different reservations in the Southwest and made different decisions regarding their heritage language and culture on their ways to successful careers. Brunn found that three factors seemed to influence these two men's decisions: (1) their parents' level of participation in traditions and levels of cultural literacy, (2) the world views and influence of the extended families and cultural groups, and (3) an individual's personal interests in traditional ways. These studies point out the necessity of addressing implied and unwritten rules and relationships as cultural discontinuity builds from procedural and interpersonal rather than the substantive issues.

Human Capital

Bourdieu (1986) defined capital "as a potential capacity to produce profits and to reproduce itself in identical or expanded form" (p. 241). According to Bourdieu, capital can be connected to three possible sources: economic (e.g., property such as a book, or a piece of land), cultural (e.g., educational qualifications acquired formally and informally), and social (e.g., social obligations or connections). Tobin (2005) stated that elements of the various sources of capital permeate the porous boundaries of cultures leading to positive or negative encounters, and therefore interpretations. Tobin & Roth (2005) emphasized the need for co-generative dialoguing, which they defined as members of a teaching team and student representatives examining specific incidents that occurred during co-teaching events. Co-generative dialoguing is meant to address and resolve emergent contradictions that might lead to the creating of negative energy and increasing cultural discontinuity that could restrict participants' ability to learn. Co-generative dialoguing can be applied in the context of PD as deep debriefing to uncover different perspectives and resolve misinterpretations, thus reducing possible negativism that could impede PD activity.

As co-facilitators and members of the L/F teams, we brought various forms of capital to these teams. The *Density and Buoyancy* immersion unit was developed by its L/F team from scratch, and team members used their sources of capital to learn together. As part of this team, Hedy felt like her sources of capital (knowledge of the science education discipline at state, nation, and international levels; expertise teaching preservice and inservice science courses and PD activities; knowledge about what happens in middle school classrooms; and professional relationships with well-known science educators) were recognized and cultivated in an effort to develop the best immersion unit in this area of the sciences. She found the co-generative dialogues during the meetings a good tool to combine her expertise with the sources of capital of others. Their sources of capital included: knowledge about classroom conditions, ELL percentages and immediate needs, implications of mainstreaming students with diverse disabilities into regular classrooms, comparisons of the results on periodic assessments performed in the district and the California science examinations, testing times, and their influence on results.

An earlier version of the *Variation and Natural Selection* immersion unit had been developed under a different grant and was tried nationally for many years prior to the efforts of this PD's L/F team. Thus, Irene joined an established team that had provided PDs on this unit for the district in the past, and whose roles were already established and rehearsed. While Irene recognized the expertise areas of her team members, team members were not aware of her sources of capital. The procedural co-generative dialoguing that occurred during PD activities did not involve renegotiation of capital such as expertise in the areas of urban teaching and preservice teaching, which represented her background. Irene had to involve her team leader in a different kind of dialogue in which she insisted on being treated as an equal to bring this to light. Both the dialogue and the way the team leader approached the issue led Irene to become more comfortable with this L/F team.

Leadership Development

Modification and integration of the work of Glickman (2002) and Ball and Cohen (as cited in Loucks-Horsley, Love, Stiles, Mundry, & Hewson, 2003) provide insight into examining the interplay between cultural discontinuity and cultural conflict in the context of leadership development in district-based large scale professional development projects (see Fig. 6.2).

Glickman (2002) developed a grid from analyzing the dynamics of two very important leadership qualities: ability to shift primary concern from self to others (horizontal axis – level of commitment) and ability to think analytically and abstractly (vertical axis – level of abstraction). As cited in Loucks-Horsley et al. (2003), Ball and Cohen conceptualized a framework that can allow the study of leadership in a classroom (the Teaching area), in a school (the Teaching Teachers area), and in a district (Teaching Leaders area).

An example of how their framework can relate to a PD project is the following: The PD takes place in the contextual milieu of Teaching Leaders (e.g., PD organizers negotiating time and activities in the PD leading to enhanced learning by the teacher participants), resulting in specific resources for the contextual milieu of

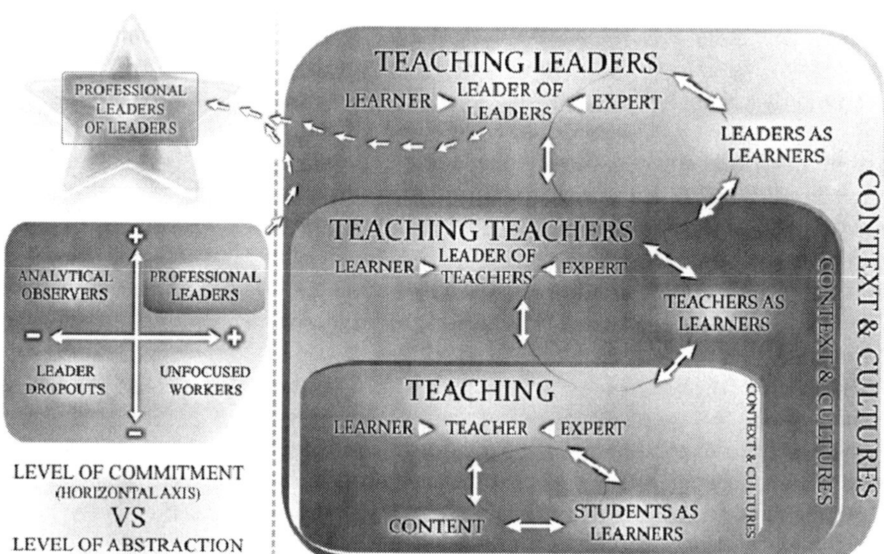

Fig. 6.2 Developing professional leaders of leaders. (Left: adapted from *Leadership for learning* by Glickman, 2002, ASCD; right: modified from Loucks-Horsley et al., 2003, Corwin Press.)

Teaching Teachers (e.g., immersion units and enhanced teacher pedagogical content knowledge) that ultimately influence Teaching (i.e., classroom teaching and student learning of science).

The combination of these ideas and works provides PD organizers an opportunity to attend to important leadership qualities in the context of the Teaching Leaders contextual milieu, thus creating a new category of individuals, Professional Leaders of Leaders. The Professional Leaders of Leaders simultaneously use their abilities to be learners, experts, and leaders. As learners they recognize others' sources of capital. As experts their sources of capital are recognized by others. As leaders they engage in practices consistent with those of individuals with a positive level of commitment and a positive level of abstraction.

This model was evident throughout our PD project as we, L/F team members, assumed learner roles and valued the capital of team members. Outcomes included the development of a higher level of trust among L/F members and needed cohesiveness in an L/F team. Irene's story is an example. Each L/F team member had capital but held different cultural beliefs about use of expert knowledge, which led to conflict and cultural discontinuity. Using her skills and demonstrating a high level of commitment, Irene made the issue explicit with the L/F team leader. Demonstrating their skills as Professional Leaders (see Glickman's grid), she and the L/F team leader resolved the conflict.

Another aspect of leadership development that guided our actions was the concepts of norms of collaboration and forms of co-presenting. Garmston and Wellman (1999) identified the norms of collaboration to include: pausing, paraphrasing, probing, putting and pulling ideas on/off the table, paying attention to self and others, presuming positive intentions, and pursuing a balance between advocacy and inquiry. Garmston and Wellman's (1999) Norms of Collaboration Inventory was initially developed to encourage positive facilitator–participants relationships and move any group toward reaching common goal(s); however, this inventory is also useful for analysis of situations related to the functioning of leadership teams. Garmston (2000) described the five forms of co-presenting, from less involved to highly involved, as: tag-team, speak and comment, speak and chart, perform-and-comment, and duet. Co-presenting using the duet form is defined as presenters in close proximity to one another; always attentive to body moves, tone, intonation, keeping eye contact; and using certain flow during the shifts from one presenter to the other.

We established group norms at the beginning of our L/F team meetings and used them to guide the conduct of team members. One example of a group norm was allowance for interruption of a speaker if listeners were experiencing and nonverbally communicating extreme confusion preventing comprehension. Duet teaming occurred during the enactment of PD activities with teachers by L/F team members sometimes finishing each other's sentences and responding to participants' faces expressing confusion.

On a daily basis during the PD sessions, we used two activities to receive immediate feedback from the PD participants about the effectiveness of the sessions. These activities, Tuning Protocol and Gots and Needs, are described in the next section.

Leadership Activities

Successful leadership activity should focus on the dynamics and relationships among members of a leadership team. Kaser, Mundry, Stiles, and Loucks-Horsley (2006) suggested the idea of team learning as this focus. They described team learning as:

> It is about getting a team to function as a whole rather than a collection of individuals. Team learning begins with a high level of self-knowledge and progresses toward developing understanding of and aligning with other team members. It encompasses the traditional goals of team building but extends far beyond. It is one of the critical factors in building communities of learners. (p. 202)

We think team learning can be achieved by (1) making explicit and working toward established goal(s) of the PD and (2) establishing and maintaining norms of collaboration among the members of the L/F teams to decrease cultural discontinuity and increase cohesiveness. In our case completing the Norms of Collaboration Inventory, Tuning Protocol, and Gots and Needs activities; and engaging in co-generative dialoguing were like keys to target cultural understanding and collaboration and help remove barriers impeding our progress. As a result we created more understanding, inclusive and cohesive functioning L/F teams, and a more enjoyable working environment.

Using the Norms of Collaboration Inventory is essential to collaborative work (see Table 6.1). Garmston and Wellman (1999) emphasized that norms must be used consistently, not sporadically. Our experience confirms how critical it is to employ consistently negotiated and established norms of collaboration in order to use everybody's expertise, make sure voices are heard and not misinterpreted (e.g., PD facilitators agreeing to use paraphrasing, probing, presuming positive intentions throughout discussions) and move meetings toward established goals (e.g., by putting and pulling ideas on and off the table). Taking care of self and others ensures that all participants' voices are heard and taken into consideration. We stress the importance of behavior becoming normative, meaning that new group members can "model their own behavior on the standards tacitly set by the veterans" (p. 37) as well as act according to explicitly agreed upon rules. We also emphasize that whenever a new team is formed within a professional development project, norms of collaboration need to be reestablished and made explicit.

A second leadership activity we propose involves daily and overall analysis of a PD experience, through use of a Tuning Protocol adapted from Blythe, Allen, & Powell (1999) and a Gots and Needs feedback process. The Tuning Protocol requires participants to focus on providing self-analysis and peer-analysis by using one warm (positive) and one cold (negative) feedback per participant. This constraint encourages participants to be clear, reflective, and supportive. Feedback from Tuning Protocols can be analyzed in light of another activity called Gots and Needs. In this activity participants write on a sticky note one thing that they have learned from the PD day (the Gots part) and, on another note one thing that they

Table 6.1 Norms of Collaboration Inventory*

Norm	Sub norm	Self-evaluate using scale below		
		Rarely	Occasionally	Frequently
1. Pausing	• Listening • Using wait time • Understanding others • No cutting			
2. Paraphrasing	• Clarification (data level) • Summarizing • Moving to concrete/abstract level			
3. Probing for specificity	• Seeking agreement/clarification • Seeking to uncover values, beliefs, assumptions			
4. Putting ideas on the table and pulling them off	• Stating goal and intentions • Revealing all relevant information • Explaining reasoning behind statements and actions • Removing/modifying own ideas			
5. Paying attention to self and others	• Aware and sensitive to own and others' thoughts and feelings • Aware of interaction patterns and group mood • Aware of moving toward proposed task			
6. Presuming positive intentions	• Assuming positive intentions • Thinking before acting			
7. Pursuing a balance between advocacy and inquiry	• Seeking equitable participation • Disagreeing respectfully and providing reasoning • Advocating politely while inquiring into ideas of others			

* Adapted from Garmston and Wellman (1999), pp. 278–279.

still struggle with (the Needs part). They place their notes on a large T-chart with one side for Gots and the other for Needs. At the end of the day, the L/F team collects and analyzes the Gots and Needs notes and the Tuning Protocols. These get addressed the following day with the PD participants. Using the Tuning Protocol and extended reflective sessions before and after each PD day can help develop team learning.

A third leadership activity is bi-weekly or monthly co-generative dialoguing (Tobin & Roth, 2005) of increasing intensity in study groups, described by Loucks-Horsley et al. (2003) as:

> Study groups are collegial, collaborative groups of problem solvers who convene to mutually examine issues of teaching and learning. They are conducted within a safe, nonjudgmental environment in which all participants engage in reflection and learning and develop a common language and vision of science and mathematics education. (p. 153)

Study groups can focus on issues pertinent to facilitation such as inquiry science, leadership models, pedagogical content knowledge (Gess-Newsome & Lederman, 1999), and making the presentations more like duets. In our case the study groups were the L/F teams.

Leadership Response

As Hedy and Irene came to the United States from different countries, elements of their cultural capital filtered through the porous boundaries (Tobin, 2005) into their L/F team settings. Due to prolonged familiarity with the current culture, Hedy felt comfortable to express her perceptions of self and her team collaborators during the Tuning Protocol and influence next day's events in light of information received using multiple sources. On the other hand, Irene felt reluctant to express her perceptions and influence the next day's agenda. This reluctance was due to her (1) expectation that her sources of capital be recognized and used by the team, and (2) experience with the leadership team leader not recognizing her expertise, ignoring her contributions, and imposing pre-determined agenda on her. Irene's reluctance and cultural discomfort led her to talk with the leader of her L/F team. This discussion led to greater team cohesiveness and changed team dynamics. During subsequent debriefings with the Tuning Protocol, the L/F team leader asked for Irene's opinion about issues under discussion, and Irene felt more valued and understood.

There is no doubt that leaders of leaders need to learn from every experience and grow constantly. Professional development L/T teams have different dynamics and cultures that need to be renegotiated with every experience to reduce the potential for cultural discontinuity. Ongoing discussion and use of the Tuning Protocol and Gots and Needs activities help professional development L/F teams to understand how home-country cultures as well as school and university cultures support and constrain interactions. Our professional interactions revealed that individuals represent different cultures who bring various sources of capital to the situation that might or might not be recognized and valued by other members. Effort needs to focus on the development of a new culture, a culture of co-facilitators who really belong to and are able to function as a team. Using co-generative dialoguing (Tobin & Roth, 2005) and the Norms of Collaboration Inventory (Garmston and Wellman, 1999) have the potential of helping individuals lead professional development projects in a cohesive way – such as the duet teams described by Garmston (2000). Understanding each others' cultures and points of reference ultimately leads to the development of the Professional Leaders of Leaders, an elegant combination/intersection between the Leaders of Leaders (as represented in the right graphic of Fig. 6.1), and the Professional Leaders (as represented in the left graphic of Fig. 6.1).

References

Bandura, A. (1969). *Principles of behavior modification*. New York: Holt, Rinehart & Winston.

Beloo, M. (2002). *"Partly Indian, partly American": Social and cultural integration of Asian Indian parents and children in the United States*. Paper presented at the annual meeting of the American Educational Research Association, New Orleans, LA, April 1–5.

Blythe, T., Allen, D., & Powell, B. S. (1999). *Looking together at student work*. New York: Teachers College Press.

Brunn, M. (1997). *Cultural interference and cultural cohesion: Schooling and tradition in two communities*. Paper presented at the annual meeting of the American Educational Research Association, Chicago, IL, March 24–28.

Bourdieu, P. (1986). The forms of capital. In J. G. Richardson (Ed.), *Handbook of theory and research for the sociology of education* (pp. 241–258). Westport, CT: Greenwood Press, Inc.

Davis, L. (1996). Equality in education: An agenda for urban schools. *Equity and Excellence in Education, 29*(1), 61–67.

Diller, D. (1999). Opening the dialogue: Using culture as a tool in teaching young African American children. *Reading Teacher, 52*(8), 820–828.

Garmston, R. J. (2000). Harmonious duos. *Journal of Staff Development, 21*(2), 65–67.

Garmston, R. J., & Wellman, B. M. (1999). *The adaptive school: A sourcebook for developing collaborative groups*. Norwood, MA: Christopher-Gordon Publishers, Inc.

Gess-Newsome, J., & Lederman, N. G. (1999). *Examining pedagogical content knowledge*. Boston, MA: Kluwer Academic Publishers.

Glickman, C. D. (2002). *Leadership for learning*. Alexandria, VA: ASCD.

Kaser, J., Mundry, S., Stiles, K. E., & Loucks-Horsley, S. (2006). *Leading every day: 124 actions for effective leadership* (2nd ed.). Thousand Oaks, CA: Corwin Press.

Loucks-Horsley, S., Love, N., Stiles, K. E., Mundry, S., & Hewson, P. W. (2003). *Designing professional development for teachers of science and mathematics* (2nd ed.). Thousand Oaks, CA: Corwin Press.

Phelan, P., Yu, H. C., & Davidson, A. L. (1994). Navigating the psychological pressures of adolescence: The voices and experiences of high school youth. *American Educational Research Journal, 31*(2), 415–447.

Sheets, R. H. (1996). *Urban classroom conflict: Student–teacher perception: Ethnic integrity, solidarity, and resistance*. Paper presented at the annual meeting of the American Educational Research Association, New York, April 8–12.

Sheets, R. (1994). *College Board Advanced Placement Spanish Literature for at-risk native speakers: A model with multicultural, bilingual, and gifted dimensions*. Paper presented at the annual meeting of the American Educational Research Association, New Orleans, LA, April 4–8.

Tobin, K. (2005). Urban science as a culturally and socially adaptive practice. In K. Tobin, R. Elmesky, & G. Seiler (Eds.), *Improving urban science education: New roles for teachers, students, and researchers* (pp. 21–42). New York: Rowman & Littlefield Publishers, Inc.

Tobin, K., & Roth, W. M. (2005). Implementing coteaching and cogenerative dialoguing in urban science education. *School Science and Mathematics, 105*(6), 313–322.

Chapter 7
Developing Leadership through Cultural Understanding in an Urban Science Community of Practice

Pamela Fraser-Abder

This chapter describes the experience of becoming a leader in an urban science education community outside my country of origin and in this process of becoming a leader, using a community of practice (CoP) to develop science teacher leaders in the urban classrooms. The chapter weaves the story of my border crossings and my CoP with the border crossings and CoPs of urban teachers. It is through these CoPs that our leadership has been fostered and enacted. The leadership activity that is found later in the chapter describes ways that have been successful for me in developing leadership in teachers. The response to that activity highlights the value of CoPs to help participants understand the complexity of border crossings.

As I traveled from my own world of a Caribbean woman of color to the often hostile world of Western science in a large urban community, I faced many personal and professional challenges. These challenges suggest the importance of understanding the issues of cultural border crossings and using this understanding to develop science teacher leaders. In this chapter leadership refers to the ability to guide, direct, and influence people. The context of my leadership is a CoP at a prestigious private urban university in a graduate education program. This CoP included preservice and inservice science teachers held together during and after their graduate program by a common program, mentoring, and social and collaborative interactions all geared toward the effective teaching and learning of science in an urban setting.

My own sensitivity to being a leader who functioned outside cultural norms and expectations, both locally and globally, enabled me to facilitate and lead the creation of a university-based CoP aimed at improving achievement in urban science classrooms. An integral part of the development of this CoP was a course, the main objectives of which were to overcome negative perceptions about other cultures, develop sensitization to issues of diversity in the K-12 science classroom, and enable participants to use this understanding to become leaders in developing CoPs in their own school setting. The course was based on the experienced and observed tensions around intercultural communications and understandings and dealt with

---

P. Fraser-Abder (✉)
Department of Teaching and Learning, New York University, New York, NY, USA
e-mail: pamela.abder@nyu.edu

K.C. Wieseman, M.H. Weinburgh (eds.), *Women's Experiences in Leadership in K-16 Science Education Communities, Becoming and Being*, DOI 10.1007/978-90-481-2239-4_7,

sensitization to issues of diversity in which participants began to communicate about their culture and develop some cultural understanding of other cultures prevalent in the urban setting. As a leader I was able to frame the structure of the CoP at my university to enable participants to develop an understanding of the importance of cultural awareness and to transport this understanding to their school communities in which they then developed CoPs in which they assumed leadership positions.

Pamela's Story: A Personal Journey

As I write this chapter and reflect on the past 17 years I realize that my leadership in this CoP evolved slowly. I did not begin with a plan to form a CoP. Intuitively I worked toward such a goal utilizing strategies which I had read about or discussed with fellow science educators. The evolution was not based on an initial literature review or the development of a research based plan. To a large extent it evolved as I responded to the need to expand the science education program which I inherited. As I worked toward the expansion of this program I used my past personal experience, my knowledge and experience in science teacher professional development, and my intuition of what would work as we attempted to teach science in large multicultural urban settings. Reflections on what led me to frame my work the way I did revealed that my decisions were grounded in the literature on community of practice and cultural border crossings.

The decision to migrate from one's country of birth brings with it many unforeseeable emotions and border crossings. Border crossings refer to both the political border crossings and also the cultural border crossings explained by Costa (1995), Aikenhead and Jegede (1999) and Giroux (1992). In 1989 I relocated to the United States with my family. Unrest in my country and the need to provide expanded educational opportunities for my children prompted this decision. Nothing in my prior experiences prepared me for the range of emotions which accompanied this move or for the survival strategies I would be forced to develop in the process of becoming American and loosing my identity as a Caribbean woman while struggling to retain my accent and cultural norms. Coming from a country in which I was always an insider into one in which I had to fit in, while being cognizant that I would always be the outsider, forced me to adjust my view of my social entitlement. Social entitlement refers to those taken for granted expectations of family and culture that would naturally presume public sphere success (Koch, 2002). In the long term, experiencing this change in social entitlement helped me to better understand the dilemmas of other immigrants who were either my students or students in the K-12 classrooms for which I was preparing science teachers.

I brought to my new setting leadership skills which I had developed with diverse international populations. These included:

- Leadership of the development and implementation of an elementary science program for the elementary school system in my native country (Fraser-Abder, 1985, 1987a)

- Professional development for science teacher educators in developing countries (Fraser-Abder, 1987b, 1988b)
- Networking opportunities for science education researchers in the Caribbean and Latin America (Fraser-Abder, 1988a, 1989)
- Leadership in the development of the International Consortium for Research in Science and Mathematics Education
- Leadership in producing a global professional development book for Science Teacher Educators. (Fraser-Abder, 2002)

My involvement in these international activities enabled me to use and develop skills in networking, intercultural communication, and sensitization to global issues of diversity in the teaching and learning of science. From working with the multiplicities of cultures and educational systems from diverse countries, I developed an understanding of the cultural norms and the often unspoken and unacknowledged biases that exist in various cultures in particular gender and cultural bias. This served as an excellent preparation for my immersion in the New York City school system and community.

I found myself in a new setting in which I was encountering migrant (international, suburban, and rural) teachers and K-12 students who, like me, were all trying to fit in and be accepted by the American urban science education community. As I began to learn about the school system, I entered classrooms where there could be more than 25 cultures represented with the teacher not belonging to, or understanding the cultural norms of, any of those cultures. I was in a city in which over 119 different languages were spoken. As my understanding of the demographics evolved, I found myself using this new understanding to revise my course and program. Because my recruitment strategies resulted in a large multicultural cohort of inservice teachers, I was able to use teaching strategies to integrate their cultural experiences into course discussion.

I was surprised by a discussion that emerged from the first assignment in this course. My students were all White. Most came from homogenous suburban or rural communities and had never been exposed to urban students. Their first assignment was to visit a classroom and observe science being taught. It was an enlightening moment for me and for them when they reported their findings. Few of them found a class in which science was being taught, and they came back with numerous reports about the conditions of the classrooms, the range of cultures represented in those classrooms and their inability to communicate with many of the students they met. They found very few minority teachers in classrooms packed with minority students.

As I continued to discover the attributes of the urban science classroom and the related teaching and student population, I realized that my success and ability to stay in this adopted country depended on my ability to guide, direct, and influence policy in my new setting. From this realization emerged my leadership in this CoP, involving a new program consisting of urban preservice and inservice science teachers.

As the program evolved, I recognized the need to increase the number of minority science teachers. I visited many schools where I saw White teachers leaving the urban setting because they found themselves unable to cope with the increasing diversity. I saw minority teachers experiencing what I was experiencing at the university level: the lack of a grounded network, a group of colleagues with whom to share a common bond and day-to-day work related problems and with whom they could brainstorm solutions. I decided to do for my students what I had been unable to do for myself – establish a network of supporting colleagues and provide for them a forum for support. I pursued a funding drive aimed at recruiting and providing free tuition for minority inservice teachers to work toward their Master's degrees and certification, and I decided to initiate a CoP at the university.

More and more as I read the literature on CoP, I realized that we were already heading in that direction. My students were doing group work, classroom observations, and action research. The preservice teachers who were all White were now in groups with the minority inservice teachers. They were meeting outside of class to complete class projects; they were going to ethnic restaurants; they were attending church together. They had started cross cultural sharing. This sharing was not limited to Black and White integration. In one group, for example, a Dominican Republic teacher was sharing with a teacher from Puerto Rico and one from Belize. We were building inter-ethnic communities within our evolving CoP comprised of White preservice teachers and inservice teachers from minority groups. The inservice teachers were guiding and slowly directing the preservice White teachers on their paths to cultural awareness and eventually cultural sensitization.

Since the process of leadership development is about the way a person broadly constructs, evolves, and enacts values, relationships, and responsibilities (Palus and Drath, 1995), it is clear that the process must take place over a span of years. To be most effective, a leadership development program helps the participants actualize their potential by providing some form of long-term support. This may consist of follow-up meetings, opportunities for participants to get together in other contexts, or a continuum of activities begun in the program. These continuing activities can foster the creation of a CoP among the participants that reinforces and stabilizes the new meaning structures that have developed (Howe and Stubbs, 2003).

With increased funding I established monthly support dinner meetings, so there was a relaxing venue where students could exchange ideas and talk about problems. An important part of these meetings was the meal. Each month the meal represented a different culture, and this provided the opportunity for CoP participants to share their culture with others. Increased funding enabled trips to conferences such as National Science Teachers Association (NSTA) and School Science and Mathematics Association (SSMA) to share what we were doing. In 1998, at the prompting of the group, I hosted our first Sharing Our Successes (SOS) conference in which graduates of the program shared the successful strategies they had developed, implemented, and evaluated in urban classrooms to increase interest, participation, and achievement in science. This conference continues to expand and welcome participants from within and outside of the state. The first conference had 150 participants and was held for one day. The most recent conference (May, 2007) was held for three days with 450 participants coming from five states.

The university forms the umbrella CoP but small satellite, school or community based CoPs have evolved. The umbrella CoP was developed in stages at the university over a period of two years and continues to evolve after seventeen years. Graduates use the umbrella CoP to maintain relationships, update their knowledge of what research says to the science teacher, share their teaching and learning stories of success or failure, and provide guidance and direction to new participants. The most recent addition to the umbrella CoP is a website of content needed to teach K-12 science. Participants, both preservice and inservice teachers, also use the umbrella CoP to keep in touch when major events arise in their lives: new job, marriage, new baby. Using the umbrella CoP as a model, graduates have developed and assumed leadership roles in CoPs in their individual schools. Interactions with the umbrella CoP sends participants back to their home school CoP reinvigorated and with new ideas for influencing their colleagues.

I have worked to develop a CoP that fosters understanding of cultural border crossings, teacher leadership, and professional development for its participants, by working closely with the state system, school systems, and school staffs, by incorporating preservice teacher education into meaningful professional learning communities at the building level, and by bringing preservice and inservice teachers together to learn from each other. Inservice teachers are involved in designing the activities of the learning community, with some members rising to leadership roles in their schools, region, and state. In the school community the professional development includes practice, feedback, reflection, and evaluation of strategies developed at the university.

Alignment with Theoretical Literature

Cultural Border Crossings

To travel between worlds is to cross cultural borders (Aikenhead and Jegede, 1999). Traveling between worlds forces one to shift from being one person in one context to being another person in a different context, while maintaining self-identity from the most familiar world. Because I experienced my own cultural border crossing, I was better able to appreciate the experiences that many students encounter when they cross borders from their life-world cultures into the culture of school science. Students experience varying degrees of success in moving between the culture of their home and the culture of the science classroom. According to Costa (1995), these transitions are smooth when the home culture and school science culture are congruent. Transitions are manageable when the cultures are somewhat different. Transitions are hazardous when the two cultures are diverse, and they are virtually impossible when the cultures are very different. Success in science courses depends on (a) the degree of cultural differences that students perceive between their life-world and their science classroom, (b) how effectively students move between their life-world culture and the culture of science or school science, and (c) the assistance students receive in making those transitions easier (Aikenhead and Jegede, 1999). In addition to directly experiencing border crossings, I have used this knowledge base in my professional development and in promoting the professional development of participants in our CoP.

Community of Practice

Communities of Practice (CoP) as described by Lave and Wenger (1991) involve learning brought about by a shared practice, social and collaborative interaction binding members together, around an area of knowledge or activity. Thus a CoP is different from a geographical community or a community of interest as it is about something rather than merely being a network of relationships. CoPs develop around activities that matter to people and lead to increased organizational capacity and capability over time, including increased number of science teachers and classroom participation in science education. Membership of any given CoP is dependent upon participation rather than position within an organization (Wenger, 1998). Participants develop partnerships whereby they engage in collaborative enquiry enabling them to learn from each other through negotiation and exchange between cultures. The key focus of such collaborative enquiry is the improvement of teaching and learning along with professional development. The CoP in my case, focused on the effective teaching and learning of science in the urban setting. As the leader of the evolving CoP and director of the university program, I used the following shared practice and social and collaborative interactions, which according to Lave and Wenger (1991), are essential to an effective CoP. All participants in the CoP were involved in a common course in which they were required to:

- Learn as much about and become sensitive to and as aware as they can be of racial, ethnic, cultural, and gender groups other than their own.
- Avoid stereotyping – never make assumptions about an individual based on their perceptions of that individual's race, ethnicity, culture, or gender.
- Get to know each student as an individual.

Participants of the CoP were encouraged to:

- Begin to assess their own conscious and subconscious biases about people who are different from themselves in race, ethnicity, culture, gender, or socio-economic status.
- Plan their curriculum within a multicultural framework while making their classroom a safe and secure haven for all students.
- Infuse multicultural instructional materials and strategies in their teaching.
- Foster collaboration and cooperation among students, parents, teachers, and administrators.

This shared practice fostered increased awareness and sensitization to cultural diversity and the challenges of border crossing. I believe that if teachers acquire these skills during their preservice or inservice program, they will be more capable of dealing with many of the problems encountered by teachers and students as they both struggle to learn and teach in their own world. Teachers have to understand that students, teachers, parents, and the community all have a role to play in achieving the goal of science for all. Each role is different but together they all contribute to the attainment of the overall goal of scientific literacy. This understanding slowly emerges by involvement and sharing in a CoP. In order to foster social and collaborative interactions participants in our CoP were required to:

- Complete group assignments which forced them to interact with each other thereby increasing their awareness of the behaviors of people who are different from themselves.
- Attend monthly meetings in which they discussed issues relevant to urban science teaching.
- Plan collaborative events, for example, attendance and presentation at conferences.

Teacher Leadership in a Community of Practice

The need for teachers to emerge as leaders in their school setting is slowly finding its way into the science education literature. In 1987 Shanker called for teachers to become leaders in their schools without leaving their classrooms. Dempsey (1992) envisioned the teacher as a fully functioning individual, a reflective practitioner, a scholar, a partner in learning, and a leader engaged with others in a shared enterprise. Katzenmeyer and Moller (1996) defined teacher leaders as those who lead within and beyond the classroom, influence others toward improved educational practice, and identify with and contribute to a community of teacher leaders.

From the perspective of Katzenmeyer and Moller (1996) leadership is viewed as arising from and created by a community of people working and growing together toward common goals. This is a view of leadership that is not managerial or administrative but is one of leadership exercised within a CoP to which the leader belongs. Teacher leaders work with colleagues within their schools, school districts, and professional organizations to introduce new ideas, support the growth of others, and lead the way toward reform. Lambert (1995) believed that leading involves facilitating the learning of colleagues in a community, and fostering the construction of meaning and knowledge together. Palus and Drath (1995) shared a model with the specific aim of developing each individual's potential for leadership within a community of practice. Their model focused on enhancing the individual's ability to participate in the leadership processes of the CoP to which he or she belongs. The three stages of their model are: (a) readiness for development, (b) the process of undergoing developmental change, and (c) a final stage in which new insights, new meanings, and new capacities for effective action have come into being. In their model, leadership development is a process that is neither linear nor unidirectional. Each step may be unstable and tentative as the individual moves back and forth between equilibrium and disequilibrium until a new perspective and new meaning structures are developed.

Professional Development and Our Community of Practice

Models of professional development of science teachers in the United States have focused primarily on content acquisition and more recently on pedagogical knowledge. According to Howe and Stubbs (2003), a broader view of professional development has only been adopted by a small number of educators who have incorporated a socio-cultural or social constructivist perspective with attention to social and personal development as well as the more traditional area of content and pedagogy. Using a social constructivist perspective of teacher development, Bell and Gilbert (1996) presented a model to change science teachers' classroom practice and their attitude toward teaching science. Their model included:

- A description of the three types of development that teachers undergo: social, personal, and professional. Social development involves learning ways to interact with others to renegotiate and reconstruct what it means to be a teacher. Personal development involves constructing, evaluating, and accepting or rejecting the new socially constructed knowledge of what it means to be a teacher and managing feelings associated with changes in activities and beliefs. Professional development occurs in phases as teachers change their practice and accept the beliefs and conceptions underlying the new activities and methods.
- The belief that development occurs within the context of a teacher-development program that includes support, feedback, and reflection.
- The absence of a prescribed time or sequence for teachers to move through the activities and phases of development.

These three aspects of the Bell and Gilbert model closely reflect the professional development of participants in the CoP described in this chapter. In the professional development for teachers in this CoP social development occurred as preservice and inservice teachers interacted with each other while jointly developing a better understanding of what is involved in science teaching and learning in the urban context. Their personal development evolved as they constructed meaning out of their interactions with each other and the urban students in their respective classrooms. Their professional development revolved around the change in their teaching practice as they reflected on the following questions: Who am I? Who are my students? And subsequently they slowly revised their teaching philosophy. The CoP provided a venue for support, feedback, and reflection through course assignments and discussion, monthly meetings and networking events. Although constrained by the length of the course, developmental issues are addressed throughout the program and extend into the mentoring received upon graduation.

Leadership Activity

Before teachers can assist students in making the transition between the home culture and the school science culture (a form of border crossing), they need to become familiar with the students' home cultures. One leadership activity which I used within our CoP is titled Walking in the Footsteps of Strangers. This activity sensitizes CoP participants to the home cultures of urban students, and engages them in reflective journaling to come to terms with tensions about cultures and to develop new understandings (Fraser-Abder, 2001, 2002). This activity represents an example of situated learning (Beach, 1999; Lave and Wenger, 1991) where the focus is on knowledge creation through participation in meaningful activities. The goal is to help CoP participants develop a better understanding of students, students' lives and home environment, and the cultural contextual framework that each student brings to the classroom.

Many culturally diverse, urban students often find themselves misplaced in our school system as they try to navigate cultural border crossings between home life and school environments. They struggle to survive in their new environment while dealing with the realities of their own life situations. It is within this context that the private reflections of teachers and students are shared. Although life circumstances are often hidden, they have a profound impact on the teaching and learning of science. This leadership activity (Fraser-Abder, 2001, 2002) focuses on the powerful cultural forces both within and outside the classroom which shape the work of teachers and reflect the tensions between their two worlds. The school often provides the forum for conflict between the teacher and student. Science teachers are often unaware of the problems faced by their culturally diverse students and also do not have teaching models for providing the required support.

This activity also provides teachers with a model for understanding cultural differences. This understanding helps them to create classroom organization and management from the inside out, instead of responding only to students' external behaviors and guessing what might be going on inside their minds and hearts (Rothstein-Fisch & Trumbull, 2008). According to Cabello and Burstein (1995) and Delgado-Gaitan (1996), teachers who are knowledgeable about the culture of school and the cultures of students' homes can serve as cultural brokers helping their students to negotiate a new cultural school environment thus becoming biculturally proficient. These teachers can also share their cultural knowledge about the home culture of their students with other school personnel and help to develop school policies that are more culturally congruent for their students. These teachers are also better prepared to use this acquired information to inform their curriculum development using strategies that are specifically geared to their culturally diverse student body.

Walking in the Footsteps of Strangers: Science Teacher Barrier Crossing Activity

When you walk in the footsteps of a stranger, you learn things you never knew you did not know. The following instructions allow you to walk in the footsteps of a stranger.

- Spend one hour interviewing a struggling student. Your interview should gather information on:
  - Student's future plans
  - Life situation
  - The influence of friends and significant others on their learning
  - Attitude and achievement in science
  - Family perceived views on education, science, future careers, jobs.
- Travel to the home area of the student paying close attention to:
  - Street environment
  - Interaction with peers, people on the street
  - Safety.
- Talk to parents at a school meeting or by telephone about:
  - Expectations for students
  - Education
  - Discipline
  - Homework rules.
- Pay special attention to TV viewing time, study area, cultural norms, gender roles, and expectations.
- What can you do? Use your journal to reflect on what you have learned about this student and develop a plan for integrating what you have discovered into developing teaching and learning strategies for that student. Remember you cannot wear their skin, because at this point in their lives it is their time to wear that skin and to deal with the pressures involved in so doing, but you can walk in their steps and begin to understand what they experience as they wear that skin.
- Prepare a 5–10 minute presentation in which you will share your reflection on what you learned about this student. DO NOT include student's name in presentation.

Reorganizing One's Thinking

The journaling and reflective presentation found in the Walking in the Footsteps of Strangers leadership activity enable individuals to be involved in what Palus and Drath (1995) described as the reorganization of an individual's thinking. The journaling and reflective presentation provide opportunities to experience five interwoven processes: experience, disequilibrium, equilibrium, construction, and potentiation.

- Experience and Disequilibrium: Individuals engage in meaning making, stretching the capacity to accept and assimilate the experience within their existing framework of reality. They experience confusion, initial lack of acceptance, or surprise at seeing their world in a new light. They are exposed to the realities of the persistent life situations of the urban child. This experience initially causes disequilibrium, and individuals can feel threatened and anxious as they resist accepting the experience as meaningful. They may not believe that this is the life experience of their footstep student. The feeling of losing one's balance is part of the Palus and Drath (1995) process. For example, one female preservice teacher's reflective daily journaling revealed her struggles with believing that there was gender bias in the science classroom. She came from a wealthy family, had attended a top rate private school, and had never experienced gender bias in her science studies. She was only convinced that gender bias existed after observing her footstep student's class over several weeks

- Equilibrium and Construction: Each individual experiences this leadership activity and the disequilibrium differently. Individuals living in the community are at a different stage of disequilibrium than those who come from homogenous white communities. Within the supportive environment of a CoP it is possible for each individual to rethink past assumptions, frame and reframe, vision and revision new possibilities, both personally and for the community of learners. Through this process, new meaning structures are built; ideas or events that did not previously make sense are now accepted. An effective leadership development program must provide an environment in which individuals are supported as they explore new avenues of understanding and eventually reestablish equilibrium. One inservice teacher who referred to himself as a "homeboy" since he attended the school in which he currently taught, his mother taught at the school and he continued to live in the community. He became our reality check when other students needed to better understand their experiences in the school and with the community and to come to terms with the reality of teaching in this setting.
- Potentiation: This term refers to the potential for future growth and development in the individual. Developmental change is never a straightforward progression through stages but involves back-and-forth movement between the old and the new until a fresh perspective is attained. Individuals who undergo the process of disequilibrium followed by the attainment of a new equilibrium become sensitized to the possibility that new perspectives and new ways of knowing can be found. They thus develop the potential for continuing development. One preservice teacher who came into teaching from a profitable career in Wall Street developed a continued friendship with the family of her footstep student. During the home visit she was invited to a meal with his family and was surprised to hear that her student had never been outside of the city. This visit resulted in his family visiting her family upstate home, and has led to a continued friendship and cultural communication between the two families. She was able to bring to the class his developing understanding of the cultural and religious norms of the family with which she was interacting.

Many teachers lack an understanding of the difficulties that culturally different students experience in their everyday lives. It is important to understand a student's background in order to provide the most appropriate learning experiences for that student. Each student has different needs that must be met. In the Walking in the Footsteps of Strangers: Science Teacher Barrier Crossing Activity, both inservice and preservice teachers were amazed at how much better they understood their footstep student, who is culturally different, on completion of the activity. Not only did they gain a more personal relationship with that student, but they also were able to relate to the student, understand the student's persistent life situation, and adapt lessons in order to help the student succeed. All our students have a story to tell. By listening to these stories teachers can better understand and support their students.

Leadership Response

Like Hodson (1992), I believe that the task of science teaching is to help all students acquire scientific knowledge, interest, skills, attitudes, and ways of thinking without doing violence to their particular cultural beliefs and experiences. Sensitizing teachers to the border crossings their students may be navigating is essential to the task of science teaching. Helping teachers develop a CoP as a support system through which they can discuss sensitive topics such as the ones that arise from an activity like Walking in the Footsteps of Strangers can contribute to a meaningful and culturally appropriate CoP. As I continue on my journey of leadership in a CoP, I have learned that:

- Teachers must be sensitive to the needs and feelings of their students. They must experience being different before they can identify with the feelings of their culturally diverse students.
- Teacher enhancement and leadership development depends on allowing them to reflect on their beliefs and practices and exposing them to alternative strategies.
- Teachers must be at the heart of the change process for the development of their leadership capacity.
- Teachers must collaborate, cooperate, and communicate about classroom interactions, instructional strategies, and relationships between school culture and learning.

The importance of community building cannot be overstated. In my story the process of community building began on the first day of the course when individuals in their first ice breaking exercise were required to talk to the person on their right for 5 minutes and then introduce that person to the group. The emphasis on working in small groups, the requirement to submit group projects, and the need to work together in order to complete assignments fostered the development of community and the CoP. This interaction is one of the necessary conditions for leadership development and precedes the development of a community that becomes a supportive network. Our teacher leaders have now taken these ways of interacting back to their schools and have created CoP in their own schools.

The program at my institution has now evolved into what Palus and Drath (1995) called a holding environment, a network of teachers who has become a community of learners. The formation of this umbrella CoP has been facilitated by several important factors. The availability of a site for meetings in a central location where teachers feel comfortable, a professional development book published annually, an ongoing monthly program of professional development in both content and pedagogy, and an annual conference have all been important in providing continuing professional development in science education, motivation, and stimulation. A five-year grant from the State Education Department and funding from private agencies have provided ongoing support for this work.

Teachers who are becoming leaders need opportunities to exercise their newly developed leadership skills. Those who had not previously been active in professional organizations have responded to encouragement, and with funding from their schools and the program have attended and presented at state, regional, and national meetings of science teachers. They have had opportunities to lead seminars, workshops, plan professional development programs for other teachers, write and field-test curriculum materials. Each of these activities has reinforced their newly acquired leadership skills and bolstered confidence. Teacher leadership is a resource that will be indispensable in making the changes that are needed in science education. The opportunity and the option to become teacher leaders should be made available to all teachers. Those who are ready will seize the opportunity to continue their own growth and development. My role as a science teacher educator and leader is to provide a focus on knowledge creation through participation in meaningful culturally sensitive activities. Social engagement provides the context for learning and transfer to take place with knowledge acquisition being inseparable from practice. The work of Lave and Wenger (1991) gives insight as to how teachers can be initiated into practice moving from the periphery toward the center, with respect to participation, in the development of contextualized knowledge, i.e. knowledge reframed with respect to the context in which they find themselves. To facilitate the formation of the umbrella CoP, I developed an approach based on collaborative enquiry, which aims to support teachers in reconstructing their own cultural knowledge and effective pedagogic practice in the urban setting in which they encounter major cultural diversity. CoP participants were encouraged to model this approach within their schools with teachers working together and using action research methods to take forward improvement-focused interventions focused on science teaching and learning.

Have I been successful as a leader of a CoP? Have I made a difference as a leader? Yes, in the lives of my students. Both my minority and White teachers leave the graduate program with a new or enhanced sensitivity to issues of gender and culture. Their experience in the umbrella CoP has enabled them to form satellite CoP in their schools, while remaining connected to the umbrella CoP, meeting monthly to discuss issues of concern and provide feedback on their practice, and annually to share with other urban educators the strategies that they have found to be successful. I recognize that I remain the focal networking point, the leader of the umbrella CoP seeking funding to maintain the efforts of the entire group and constantly learning about what works in urban professional development of science teachers.

Did I sacrifice my rise as a leader in academia to achieve this goal? In research universities professional development is not as valued and as rewarded as quantitative and qualitative research, although it is sometimes possible to combine both. Given funding restrictions and the administrative responsibilities inherent in such work, combination is usually not feasible. In the process of deciding to become involved in professional development work, one needs to be aware that higher education does not currently value, in terms of tenure and promotion, the decision to pursue such a path. This means that one could potentially sacrifice one's climb up the tenure and promotion ladder. Certainly I could have accepted a job elsewhere where the climate was more conducive to my upward journey in academia, but I chose to stay and help others like me move into positions where they can help our urban K-12 students. I have been responsible for making a significant contribution to the number of minority teachers in the urban science classroom. My influence has spread beyond the umbrella CoP. In a recent presentation at a national conference I had in the audience a middle school teacher who 10 years ago had heard me talk about the footstep activity in this chapter and was so excited about it that she returned to her school district and successfully implemented the activity. Her presence at my session with one of her students was testimony of my ability to guide, direct, and influence people. My best day is one in which one of my students drops by the office or calls to say that he or she is still using some strategy which evolved from one of my courses or from our CoP.

I realized that I too have found a home in this CoP.

References

Aikenhead, G. S., & Jegede, O. J. (1999). Cross-cultural science education: A cognitive explanation of a cultural phenomenon. *Journal of Research in Science Teaching, 36*(3), 269–287.

Beach, K. (1999). Consequential transitions: A socio-cultural expedition beyond transfer in education. *Review of Research in Education, 24*, 101–139.

Bell, B., & Gilbert, J. (1996). *Teacher development: A model from science education*. London: Falmer.

Cabello, B., & Burstein, N. (1995). Examining teachers beliefs about teaching in culturally diverse classrooms. *Journal of Teacher Education, 46*(4), 285–294.

Costa, V. B. (1995). When science is "another world": Relationships between worlds of family, friends, school and science. *Science Education, 79*, 313–333.

Dempsey, R. (1992).Teachers as leaders: Towards a conceptual framework. *Teaching Education, 5*(1), 113–120.

Delgado-Gaitan, C. (1996). *Protean literacy: Extending the discourse on empowerment*. Washington, DC: Falmer Press.

Fraser-Abder, P. (1985). *Development, implementation and evaluation of the science – A process approach for Trinidad and Tobago [SAPATT] science curriculum*. (ERIC Document Reproduction Service No. ED 258 834)

Fraser-Abder, P. (1987a). The development of primary science education in Trinidad and Tobago. *Caribbean Curriculum, 1*(1), 55–67.

Fraser-Abder, P. (1987b). *Primary science teaching training project*. Liverpool, England: Commonwealth Secretariat/UNESCO.

Fraser-Abder, P. (1988a). *Sourcebook of science education research in the Caribbean*. (Science and Technology Education Document Series, No. 26). Paris, France: UNESCO.

Fraser-Abder, P. (1988b). *Primary science teacher training for process based learning*. Barbados: Commonwealth Secretariat/UNESCO.

Fraser-Abder, P. (1989). Science education research in the Caribbean: Retrospect and prospect. *Studies in Science Education, 16*, 219–227.

Fraser-Abder, P. (2001). Preparing science teachers for culturally diverse classrooms. *Journal of Science Teacher Education, 12*(3), 287–313.

Fraser-Abder, P. (2002). *Professional development of science teachers: Local insights with lessons for the global community.* New York: Rutledge Falmer.

Giroux, H. A. (1992). *Border crossings: Cultural workers and the politics of education*. Minneapolis: University of Minnesota Press.

Hodson, D. (1992). Towards a framework for multicultural education. *Curriculum, 13*, 15–18.

Howe, A. C., & Stubbs. H. S. (2003). From science teacher to teacher leader: Leadership development as meaning making in a community of practice. *Science Education, 87*, 281–297.

Katzenmeyer, M., & Moller, G. (1996). *Awakening the sleeping giant: Leadership development for teachers*. Thousand Oaks, CA: Corwin Press.

Koch, J. (2002). Coping with feelings of fraudulence. In J. Cooper & D. Stevens (Eds.), *Tenure in the sacred grove* (pp. 107–116). Albany, NY: SUNY Press.

Lambert, L. (1995). Toward a theory of constructivist leadership. In L. Lambert, D. Walker, D. Zimmerman, J. Cooper, M. Lambert, M. Gardner, & P. Slack (Eds.), *The constructivist leader* (pp. 28–51). New York: Teachers College Press.

Lave, J., & Wenger, E. (1991). *Situated learning: Legitimate peripheral participation*. Cambridge: Cambridge University Press.

Palus, C., & Drath, W. (1995). *Evolving leaders: A model for promoting leadership development in programs*. Greensboro, NC: Center for Creative Leadership.

Rothstein-Fisch, C., & Trumbull, E. (2008). *Managing diverse classrooms: How to build on students' cultural strengths*. Alexandria, VA: Association for Supervision and Curriculum Developments.

Shanker, A. (1987). Teachers as school leaders. *American Teacher, 71*, 5.

Wenger, E. (1998). *Communities of practice: Learning, meaning and identity*. Cambridge: Cambridge University Press.

Chapter 8
Wither Thou Goest: The Trailing Spouse or Commuter Marriage Dilemma

Roxanne Greitz Miller

Entreat me not to leave thee, or to return from following after thee; for whither thou goest, I will go; and where thou lodgest, I will lodge: thy people shall be my people, and thy God my God; Where thou diest, will I die, and there will I be buried. The Lord do so to me, and more also, if aught but death part thee and me.

~ *Ruth 1:16–17, King James Bible*

The Bible verse above is often used during wedding ceremonies to capture the intended devotion of spouses to each other. However, over the last several years of my 16-year marriage, my spouse and I have faced the modern reality that, when both partners are highly invested in their careers and occupy leadership roles, it is often difficult to find promising employment opportunities for both partners in the same geographic location. During our marriage, we have lived together as a traditional couple and family, sharing a house in the same city 365 days a year. We also have each occupied the role of trailing spouse, whereby, to continue living as a traditional couple/family, one of us has made a career sacrifice for the other. Currently, we are partners in a long-distance commuter marriage in which we each live in our own house, 2,712 miles from each other, for three weeks of every month.

I realize that my relationship with my spouse may be different from many other marriages or partnerships. Both my spouse and I are very independent and self-reliant people. We believe that both partners in the relationship should have equal voice and participate in shared decision-making, but we also believe that partners must sometimes sacrifice for the other when confronted with situations when compromise is not possible. We also both devote considerable attention to the direction of our careers and our career goals, particularly with respect to what we wish to accomplish by the end of our careers. We both embrace a scientific mindset when we approach problems of any type, even domestic: doing extensive research, gathering data, formulating hypotheses, and drawing conclusions that may be revised later in light of new evidence. Knowing this about our individual personalities and

R.G. Miller (✉)
School of Education, Chapman University, Orange, CA, USA
e-mail: rgmiller@chapman.edu

K.C. Wieseman, M.H. Weinburgh (eds.), *Women's Experiences in Leadership in K-16 Science Education Communities, Becoming and Being*, DOI 10.1007/978-90-481-2239-4_8,

the way we navigate through our lives individually and as a couple is integral to understanding our decisions.

Currently, between 700,000 and one million couples in the United States alone are involved in long-distance relationships of some type (Guldner, 2006). Although the concepts of trailing spouses, commuter marriages, and long-distance relationships are far from new, there is little discussion in the related literature on how those in career leadership roles make their decision whether to live as a traditional couple/family, to have a trailing spouse, or to embark on a commuter marriage or long-distance relationship (Guldner, 2006). Additionally, while the literature often includes examples of dual-academic career couples (Achurperil, 2003; Guldner, 2006; Mannix, 2001; Villalba, 1999a, 1999b; von Ruschkowski, 2003; Wilson, 1999), I have found no references to dual-career couples who straddle the academic and corporate world, as do my husband and I. Through my own story, I will address how certain principles of leadership have affected my decisions and those of my partner, as we have lived through all of these situations. Independence, resources, flexibility, and communication have been prerequisites for our being able to sustain our relationship and support our goals. I also will discuss the effects of these decisions on both my professional and personal life.

Before beginning my story, I must say that we realize the decisions we have made would not, and could not, be made by many others, and we honor those individual differences. This piece should not be interpreted as representing our actions and situation as the only, correct, or best path to take – this is simply how my spouse and I have handled our own personal situation.

My Story: From Public School Science Teacher to Tenure Track Professor

In *Tenure in the Sacred Grove: Issues and Strategies for Women and Minority Faculty*, Collay (2002) stated, "No faculty woman can ignore the complex interweaving of family and work as she makes meaning of her career" (p. 91). In keeping with this, I begin with the story of my education and career. I was the first, and only, person in my family's multi-generational household to attend college, made possible for me through financial scholarship support. After completing my bachelor's degree in biology and secondary education, I was a full-time secondary science teacher in the southeast United States, achieving recognition at local and national levels for my science teaching. Because I wanted to continuously improve my teaching and to have additional career options, such as working in science teacher education at the university level, I attended graduate school at night during the public school year and full-time in the summers. This enabled me to complete my master's and doctorate in science education during my first seven years of public school teaching. Also during that time, I married an organic chemist whose family's educational background was very different from my own; all of his immediate and most of his extended family were college educated. The son of a career academic in physics and a recent Ph.D. himself, my husband was well acquainted with university life and the rigors of graduate school, in which I was enrolled when we met and married.

Since leaving graduate school, however, my husband has worked exclusively in corporate scientific industry, not academia, rising to the level of vice president in charge of research for two of the companies in which he has been employed. As such, he is very much a leader amongst his peers. A long-standing joke between us was that he had 18 publications before finishing graduate school and that I had to catch up, which I did by my second year on the tenure track.

As Booker (2003) noted in the *Chronicle of Higher Education*, while completing a doctorate in a field other than education in your twenties is commonplace, completing a doctorate in education at such a young age is unusual. Indeed, most of my fellow graduate students were 10–20 years my senior. Questions from a select few of my male professors, whose spouses had assumed the career role of professional wife, ranged from why I worked so hard by taking a full load every term while teaching, rather than commending me for being able to do so, to why I was getting this degree if I planned to have a child soon after. The high dropout rate among the few young women with children in my Research I institution's graduate programs likely fueled their beliefs. Looking back, I wonder whether the questions were more benevolent than I had thought at the time, reflecting the realities of academic job opportunities for a married mother of a school-aged child whose father's salary was as much as three times my own.

In response to the difficult demands I saw facing my fellow female graduate students with young children, I developed with my husband's input and support what became known as my career Master Plan. In it, I aimed to complete my doctorate before having a child and then to move on to university-level science teacher education and research, with the expectation that my recent experience in the public school classroom would provide me with perspective on the day-to-day realities faced by teacher education candidates. I optimistically planned for my child's infancy and toddler years to include flextime positions, such as adjunct lecturing and research, which would prepare me for the tenure-track and bridge my secondary teaching experience to university work. I was confident that, while there would be inevitable struggles, I would be able to successfully handle being a mother and an academic and that my tenure-track goal was attainable. If I were to find that it were not, I would be happy to return to the public school classroom as a science teacher, enriched by the experience I had gained.

I received my doctorate shortly after my 29th birthday, and continued to teach science and supervise teachers in my public school through concurrent administrative positions as department chairperson and coordinator for exceptional education. About a year after graduation, I became pregnant, with everything going according to the Master Plan. Then, early in my pregnancy, my husband received a promotion that included a cross-country transfer to the West Coast U.S. corporate headquarters. For my husband, it was a choice between a promotion or eventual unemployment, as the smaller regional division of his company on the East Coast would eventually be closed. With a baby on the way and my public school salary too low to support all three of us, my husband accepted the promotion. He lived in a hotel for several months, while I stayed behind and continued to teach during the remainder of my pregnancy. In this situation, our decisions were based solely on the economic reality that my husband's continued income was necessary for our growing family. We thus began our first experience with a commuter marriage, with my husband traveling back for a weekend visit once a month throughout the remaining six months of my pregnancy. After our son was born, I resigned my public school position, and our son and I moved across the country to join my husband. Thus, at that time, I was a trailing spouse, unemployed for the first time since age 16, with no job prospects on the horizon.

After settling into motherhood and our new home, I found, serendipitously, that there were a number of universities within an hour's commute. About six months after we arrived, my husband read in the local newspaper that the closest Research I university received a very large federal grant in reading and writing in the science content area, and he encouraged me to contact the university. I called the principal investigators to offer my experience in science education and inquired as to whether I could apply for a position on the team. I was hired, and a postdoctoral scholar position was created for me out of the grant funds. When I told my husband how lucky I felt, he told me that it wasn't luck. Rather, it was the result of my hard work and experience, as well as my making my own opportunities.

My faculty sponsor was the Dean of the Graduate School of Education, and after the first full year of the project, he asked me to assume the role of co-principal investigator. This made me the person responsible for the day-to-day operations of the grant and, in the process, gave me a history of funding. He also introduced me to his colleagues across the country and encouraged me to make the transition into an educational leader who could write both publications and grant applications. As a result of the continuation of funding of the grant, my postdoctoral appointment stretched from three to five years, and I had the opportunity to publish in diverse areas, present at numerous top-level conferences, and network with colleagues across the country. It was a dream job for that stage in my life. Like another "mommy candidate" (Goode, 2004), I've often been asked how I managed to do all this with a preschooler at home. My answer is the same as Goode's: it wasn't easy, but I just did it. It was part of my Master Plan. Confident of my ability to get a local position if one were available once my postdoctoral position concluded, my husband and I decided that my first tenure-track search would coincide with our son's entrance to kindergarten.

Circumstances that year were such that positions in secondary and science education were available at several institutions to which I could commute. Thus, after spending most of our son's year in kindergarten searching for a position, I signed a contract for a tenure-track position at a private master's comprehensive university 20 miles west of our home. I read every piece of literature I could find on the search and negotiation processes. I found *Women Don't Ask: Negotiation and the Gender Divide* (Babcock & Laschever, 2003) to be essential and to make a big difference in the results I achieved. I negotiated a starting salary that was 10% over the initial offer at my first choice institution, with limited class preparation for the first two years and seed money for the first three years. Shortly after signing the contract, I received a significant amount of federal grant funding, enabling me to start my first year as a university faculty member in a strong position for moving toward tenure. Then, the inevitable happened. Two months after signing my contract, before I started my tenure-track position, my husband received an excellent job offer with a different company that would once again advance him to a scientific leadership capacity. This job, however, was back on the East Coast.

We struggled to make a decision. With the last job offer my husband had been the financial leader and it was a natural time for a transition in my life, ultimately leading to my assuming a new position of leadership in science education. Now, I was at the beginning of a promising appointment, full of possibilities, but was clearly not the financial leader of the pair. Had we still been a childless couple, I would have likely told him to take the position and for us to again engage in a commuter marriage for an extended period of time. With a young son, however, I did not want either of them to suffer from the separation. Additionally, because I had not yet started my new position as a university faculty member, I could not predict whether I would be able to manage the stresses of my first years on the tenure track while raising our son essentially on my own. Our leadership in our careers was now challenged by our need to be leaders in our family, setting worthy examples for our son, while taking into consideration both the financial and career implications of our decisions.

For these and other reasons, we opted for my husband to refuse the offer. This has brought us to the notion that the term trailing spouse should be expanded to apply not only to those who follow their partners, but also to those who give up opportunities in favor of the other. For us, although no one changed jobs, my husband became the trailing spouse in our partnership. His giving up an opportunity alone was enough to qualify him as trailing.

My husband continued to work for his employer on the West Coast for two more years, while still being courted by the company whose first offer he had refused. During this time, I adjusted to academia, received promising internal reviews, and secured additional grant funding. After several offers, counteroffers, and endless discussions between us and with the East Coast company about our options, we decided that my husband would sign a 21-month employment contract and take the position as I neared the end of my second year on the tenure-track. Although I did not want to leave my tenure-track position, considering that half of my shortened four-year tenure clock had elapsed, I investigated employment options in this new location, but none was comparable to what I had. I would have been relegated to a part-time adjunct faculty member, at best, so we decided that I would continue in my tenure-track position and that we would engage again in a bicoastal commuter marriage, but this time as a family. Part of my husband's contract allows him to return to California for one week a month to visit, but we do so at our financial expense. Unlike our prior short-term experience as a commuter couple, where his former company paid for the one weekend a month trip home, my husband's current employer does not pay for any costs related to his maintaining a second home or his travel between our homes. Thus, our decision to engage in this bicoastal living arrangement is a substantial drain on our financial resources, but one that we hope will be temporary and recoverable in the long run if I obtain tenure and if the company my spouse now works for is successful in their new ventures. Even if one, or both, of these outcomes is negative, we both currently believe that we made the right decision to take these risks, given the potential long-term returns for each of our careers.

As I reflect on my journey to date, it becomes apparent what a complex social and economic system the dual-career marriage with children becomes when both partners are leaders in their fields and cannot work for the same institution or even in the same location. From my experience, I believe that it is more difficult for those seeking tenure/tenure-track positions in academia when relocating to find employment than for those in business or for those in public K-12 schools. First, universities are generally located in relatively urban areas placing geographic limitations on the possibilities. Second, universities typically have only a limited number of open tenure lines per year. As such, they are generally unable to immediately create a tenured/tenure-track position to accommodate someone. Spousal hires at the university are usually reserved for spouses of people being hired by the university, not for spouses of those being hired into local industry outside of the university. Additionally, due to the limits on the number of available tenure lines, spousal hires are often not tenure-track positions. Third, and of particular importance to those of us in education, as opposed to other academic disciplines, is that if we are fortunate enough to find a position at another university in our new town, we must begin anew, building ties to school systems and teachers, with no guarantee of success. This presents a labyrinth of challenges for one who is at a certain level of leadership in science education and has grants, programs, and/or lines of research in place.

How Leadership Theory Has Influenced My Story

Our marriage and our major decision-making related to our careers have been influenced by both situational leadership theory (Blanchard, Zigarmi, & Zigarmi, 1985; Hersey, 1985) and transformational leadership theory (Downton, 1973). Situational leaders choose the best course of action based on the situation and adapt their leadership styles to each situation with which they are faced. Transformational leaders engage with others through enactment of a shared vision so that leaders and followers are both raised to higher levels of motivation and morality, focusing not on short-term goals, but on higher-order intrinsic needs. Using leadership theory to help craft our common goals has given my husband and me similar vocabulary to use when talking with each other about our ideas, enabled us to see each others' point of view through different lenses, and inspired us to reach farther in our career goals than we might have seen possible. We often read the same books and articles that were found by one of us and share them with each other for discussion. Several of these readings are featured in this chapter.

One of my favorite authors on leadership and management is Marcus Buckingham who, as a former consultant for the Gallup Organization, has had the opportunity to interview countless people in myriad careers about what they believe is essential to good leadership and management. I raise the issue of management here because I believe that it is essential to have a working definition of a leader and a manager (which are not synonymous) and to examine how faculty members function within both of these roles. In academia, leaders and managers perform roles that I believe are quite different from those performed in business settings to which leadership theory is traditionally applied.

According to Buckingham (2005), in *The One Thing You Need to Know* and *First, Break All the Rules* (written with Coffman, 1999), great leaders are those who: (1) rally people to a better future, (2) have extreme clarity of vision, (3) are able to represent that vision to others and (4) are able to motivate them toward achievement. I believe that, in academia, we often use the term leader to represent a person of extreme professional productivity, rather than necessarily one of vision and motivational skills. For example, a faculty member who publishes at an outstanding rate and possesses a large amount of grant funding would be likely to be referred to as a leader in his/her field, but would that person necessarily fit the description of leader found above? Possibly not. As such, we need to be mindful that when we, as academics, examine leadership theory, our connotation of leader may not concur with that of others.

A commonly quoted adage about managers and leaders is that managers manage complexity, while leaders manage change (Kotter, 1990). This statement describes the leader's role as a visionary and change agent, consistent with Buckingham's conceptualization. Managers, in contrast, manage complexity, with the goal of making other people more productive and successful. They problem-solve situations, organize tasks, plan budget, and respond to multiple demands. I propose that university faculty members take on a combined role of leader–manager in almost everything they do, unlike in business where these roles are most often separated. If a faculty member makes a great contribution to the field, then it is likely that he or she possesses a vision of the field in which the work is done, consistent with the Buckingham definition of a leader. However, at the same time, that leader is also assuming the role of manager of his or her career. The outstanding contributions to the field cannot be made without the faculty member managing a high degree of complexity, attention to detail, planning, and budgeting of time and resources. Unlike leaders in business, most faculty members do not have managers under them to carry out these tasks, especially at smaller institutions like mine where graduate or doctoral student assistants and faculty support staff members are limited. We occupy this role of manager ourselves within our own personal careers. Even though we have department chairs and deans who oversee our work, we are primarily self-managers and self-motivators.

It is no different for my husband and me in our marriage. We are each occupying the combined role of leader–manager as well. We manage ourselves, our behavior, and our choices as a couple, while also acting as leaders who rally for each other as we manage the complexity of our bicoastal family life. Together we serve as monitors for the other, evaluating our actions and situations and gauging our degrees of success and failure.

Leadership Activities and Responses

In this section, I illustrate key issues to bear in mind if you are negotiating the combined leader–manager role in academia, and within your personal relationship. Using my trailing spouse and commuter marriage experiences, I have summarized the issues below and then have italicized them in the text. The key issues are:

- Envisioning and articulating desired career achievement with one's spouse
- Comparing opportunities to the career vision, in my case, the Master Plan, and weighing options against goals
- Using resources (e.g., *First, Break All the Rules, Now, Discover Your Strengths, Go Put Your Strengths to Work, The One Thing You Need to Know, Women Don't Ask, Tenure in the Sacred Grove*) to maximize productivity and success, as well as applying suggestions or questions to:
    - Evaluate desired and actual workplace strengths
    - Guide job search and negotiation, in view of career vision
    - Situate professional decisions in alignment with personal relationships
- Identifying core strengths and using them in professional and family life contexts or know who you are and what you need
- Dividing responsibilities in the relationship and family, as well as drawing on strengths and what each likes doing as a means to nurture relationship
- Being ready for the unexpected
- Making the best decisions for careers, relationship, and family out of devotion to the relationship and a shared family future.

If we adhere to Buckingham's definition of a leader, we need to understand how vision has been important to the leadership decisions that my spouse and I have made. We have dedicated a great deal of time to *envisioning* what we want to achieve in our careers and where we want to ultimately be when our careers end. We also *articulate* our vision to each other. My Master Plan, which charted the progress I wanted to make during the past 18 years of my career, is evidence of this vision. When unexpected situations, such as my husband's promotion and transfer, arose, we both *examined such situations in view of the Master Plan and weighed our options against our goals*. I want to add that my husband also has a career Master Plan, but I have intentionally not shared it here, as it is his story to tell. We have never veered from the vision we created together. Instead, on a case-by-case basis, we have made decisions based on opportunities and challenges and have motivated each other toward achieving our goals in line with transformational leadership principles. We have motivated each other toward higher levels of achievement that addressed our own intrinsic needs, while maximizing our contributions to ourselves, to our family, and to the communities in which we live.

While there are many principles for university faculty members to learn as a means to *maximize productivity and success*, I have concentrated on the following throughout my career. Going into my tenure-track search, I knew that I wanted to join a strong School of Education, one in which young scholars would be given the opportunity and support to enable them to grow. In trying to empirically establish what such a School of Education might look like, I encountered Buckingham and Coffman's (1999) *First, Break All the Rules*. In it, they list 12 questions that they used during their work for Gallup interviewing thousands of people in different types of workplaces (including some schools). They found that positive answers to these particular 12 questions yielded the most conclusive *evidence of a strong workplace* and one where I would find what I needed. These questions are:

1. Do I know what is expected of me at work?
2. Do I have the materials and equipment I need to do my work right?
3. At work, do I have the opportunity to do what I do best every day?
4. In the last seven days, have I received recognition or praise for doing good work?
5. Does my supervisor, or someone at work, seem to care about me as a person?
6. Is there someone at work who encourages my development?
7. At work, do my opinions seem to count?
8. Does the mission/purpose of my company [department, School, College] make me feel my job is important?
9. Are my coworkers committed to doing quality work?
10. Do I have a best friend at work?
11. In the last six months, has someone at work talked to me about my progress?
12. This last year, have I had opportunities at work to learn and grow? (p. 28)

I *used these questions as criteria during my job search* for my tenure track position. Overall, they served me well as a guide for evaluating the comments made during interviews, for estimating my relationships with potential colleagues, and also for reviewing all of the documents and established university procedures that are presented to the candidate during the search process. Even so, some of the 12 questions are more germane to institutions of higher education than are others, and some questions could not be answered definitively during the search and would have to be answered after taking the position. For example, Question 4 is less applicable in an academic environment in which faculty are loosely supervised and could not be answered during the search itself. Part of my formal *contract negotiation* with my dean was based on Questions 1 and 2. Specifically, I wanted to make sure that I knew what was expected of me in terms of number of classes, course preparations in the first three years, and research/productivity demands. I also wanted to be confident that I would be able to obtain the materials and equipment to do my work well. This meant that, upon my arrival, I would be able to obtain seed money in the event that I was between funded grant projects. Knowing that these particulars would not be reflected in the standard faculty contract, all of my negotiations were done via email so that, in the event that I needed documentation, I would have it. I later found out that this was highly unusual, compared to my colleagues' experiences in departments, schools, and colleges of education across the country. Most had no limits on teaching demands or were given no seed money.

I also found Questions 9 and 10, particularly Question 10, to be very important. In general, I am highly independent, so for me to highlight these two social dynamics questions merits attention. As a woman with a young child, I wanted to be able to have some female colleagues who were roughly my age and who were balancing the same family demands as I was. I knew that I needed colleagues, male or female, who not only would understand my struggles, but also with whom I could share the experience. I crossed two institutions off my list because I found, during the interviews and my presentations of research, that the faculty members were highly insular and there was a distinct lack of colleagues with young families. Although the departments were family friendly, the potential to form deep friendships with other women, based on common experiences, was limited. I now have a best friend at work, whose daughter is a few years younger than my son, and we have gotten to know each other through the personal and professional demands that we have in common. In addition, my colleagues in the School of Education are friendly, helpful, and supportive, which I believe contributes greatly to my work.

Prior to writing this chapter, I had never *considered the 12 questions with respect to a personal relationship*, but I now feel that they are very much applicable to the commuter marriage in which I now find myself. My husband and I constantly make sure we understand and discuss our roles in this living arrangement, knowing what is expected from each of us when we are apart and when we are together, as well as making sure that we have the tools to do our jobs. Our tools have been creative and include frequent email, no-limit cell phones for everyone in the family, and dedicated funds to support ourselves in this situation. We find that communication, praise, encouragement, and feedback are even more important now than they were when we lived in the same household. They are essential to keep us functioning in view of the demands and stress we face.

The large body of literature on long-distance relationships (Guldner, 2006) contains a recommendation that the partners in the relationship sit down together every six months to talk openly about the relationship and evaluate it. I find this to be similar to the purpose of question 11. Questions 6 through 9 focus on couples supporting each other, maintaining high levels of contribution to the relationship and to the mission of the bicoastal situation, and valuing each other's opinions about both the situation and each other's work. Fortunately, my husband and I discuss with each other not only the broad strokes of our professional lives, but also the minutiae. This helps us to remain connected, and we learn from each other's experiences, even though we are in different fields. Interestingly, it was my husband who first introduced me to many of the leadership and management books that I have read, and I send him readings on educational theory and practice, which he finds useful as well.

Because I am trying to sustain my contribution to my field and to my marriage, I need to assume the combined leader–manager–contributor role on a daily basis. In this regard, I am turning to Buckingham's latest works, *Now, Discover Your Strengths* (Buckingham & Clifton, 2001) and *Go Put Your Strengths to Work* (Buckingham, 2007) for advice. In *The One Thing You Need to Know*, referenced earlier, sustained individual success was portrayed as following one principle: "discover what you don't like doing and stop doing it" (p. 217), through varied applications. Out of this principle has grown the current strengths movement in leadership and management theory, whereby people *identify* their *core strengths and concentrate on using them,* rather than trying to instill strengths that are not already present in a person. This is an oversimplification, but the idea rings true to me. Although I had not read *The One Thing You Need to Know* at the time that I decided to take on university work, I had concluded, through my experiences as a teacher in the public schools, that I did not like administration and that there was going to be very little opportunity for me to advance and engage in new opportunities in my public school teaching career without taking on administrative roles. Thus, my leaving the public school classroom for higher education was, in a way, ceasing to do something that I did not like – administration – and diverting myself to what I do like and to what I am good at, which is teaching and research.

In general, I try to apply the principle in my professional life wherever I can. As mentioned earlier, I am naturally a scientific person, who likes to think in an experimental manner. As such, working in research is a good fit and an effective use of my time. I do not like committee work or many other administrative tasks in which faculty members often find themselves engaged. This does not mean that I am not good at these tasks. In this regard, strengths theory dispels the notion that if something is not one of your strengths, then it is a weakness and something that you do not do well. To the contrary, one can perform extraordinarily well in areas that are not strengths. In fact, people often choose vocations that they can do well, but do not like, in the interest of financial gain or career development. This is what my spouse did for a large part of his career and at his former company. The downside to this approach is that working in areas that are not your strengths draws time and energy from activities that speak to your inner self. Being more engaged in activities that spoke to his strengths, and less engaged in those that he did not like, was a major motivator for my husband to take his current position and a major reason why I supported his decision, despite the obstacles presented by the commuter marriage it involves. Simply put, we both believe that there is great value in people being employed in positions that allow them to do what they like and are able to do well.

In a bicoastal family, with each of us on separate coasts and maintaining our own houses, there is little room to ignore activities that are not our personal strengths on the domestic front. When you are the only person available to do certain tasks, you do them. However, we are trying to *divide* our responsibilities around our shared home, *draw on our strengths and what we like to do*, and minimize the time we spend on those things that are not our strengths. I can definitely say that raising our son without my husband present 75% of time has been a challenge, but it has also restored a focus on my family that I lost during my first two years on the tenure-track. Then, I scrambled to do what seemed like everything all the time and I allowed work demands to intrude on my personal life at almost any opportunity. Now, when I am at home with our son, he is the center of my attention. My weekends and the week each month when my husband is at home are times I set aside and keep more sacred for family activities, and I am grateful to have had this time and experience to help me gain more balance in my life. My husband has experienced a similar positive change. We joke that we have more quality time together in our one week a month than when he lived in the same house, and we enjoy our time together more because he is happier now that he is engaged in work that inspires him and is tailored to his strengths. And, for those who are wondering, our son is doing very well and seems to be relatively unaffected by our change in family status. He is old enough to understand that this situation is temporary, that our family is as stable now as it was before, and that if we found that he were negatively affected that we would seek another living situation as soon as possible.

Where will we go from here? We have been a bicoastal commuter family now for nearly one year. Our son is now nine years old and will finish the third grade in four months. My husband's employment contract should end very close to the time that I receive my tenure decision next academic year. If granted tenure, our current plan is that my husband will return to California and to our shared home, and we will apply all that we have learned in leadership and management theory to a new job search for him. At the level of leadership in which he finds himself, searches for scientific officer positions typically last from many months to more than a year, similar to university faculty searches. However, if my tenure decision is negative, or if my husband is unable to find employment back on the West Coast and my income as a tenured professor alone is not sufficient to sustain our family, we may both be searching for new positions elsewhere. We realize that once again one of us may find ourselves occupying the role of trailing spouse in order for all of us to live together as a family. In any case, we know that *the unexpected, either positive or negative, can arise at any time* for either of us, and we will continue to *make the best decisions that we can for our careers, for ourselves, and for our family.* Our *devotion* to each other is not represented by the willingness to go wherever one partner's career leads. Rather, it is embodied by our willingness to support each other on our individual paths and through the trials that accompany them on the greater road toward our *family's shared future*.

For others who are facing the same situation, I think the best advice I can give is this. First, it has been statistically established that long-distance relationships succeed and fail in the same proportions as do traditional relationships (Guldner, 2006). One does not necessarily have to choose between the relationship and the job; nor does choosing a long-distance relationship necessarily jeopardize the relationship's success or the family's stability. Therefore, assuming professional leadership roles while being involved in a dual-career couple, with or without children, is indeed possible for both partners. However, it takes independence, resources, flexibility, and communication for the long-distance relationship to be successful. I believe that the inner strength gained from my experience in our long-distance relationship at this critical time in my academic career is a benefit I will draw upon for the rest of my life.

References

Achurperil, M. (2003). Married, living apart, still hopeful. *Science*. Retrieved December 1, 2007, from http://sciencecareers.sciencemag.org/career_development/previous_issues/articles/2240/married_living_apart_still_hopeful/

Babcock, L., & Laschever, S. (2003). *Women don't ask: Negotiation and the gender divide.* Princeton, NJ: Princeton University Press.

Blanchard, K., Zigarmi, P., & Zigarmi, D. (1985). *Leadership and the one-minute manager: Increasing effectiveness through situational leadership.* New York: Morrow.

Booker, K. (2003, August 5). The disadvantages of youth. *Chronicle of Higher Education Online.* Retrieved September 17, 2007, from http://chronicle.com/jobs/2003/08/2003080501c.htm

Buckingham, M., & Coffman, C. (1999). *First break all the rules: What the world's greatest managers do differently.* New York: Simon & Schuster.

Buckingham, M., & Clifton, D. O. (2001). *Now discover your strengths.* New York: Free Press.

Buckingham, M. (2005). *The one thing you need to know ... about great managing, great leading, and sustained individual success.* New York: Free Press.

Buckingham, M. (2007). *Go, put your strengths to work: 6 powerful steps to achieve outstanding performance.* New York: Free Press.

Collay, M. (2002). Balancing work and family. In J. E. Cooper & D. D. Stevens (Eds.), *Tenure in the sacred grove: Issues and strategies for women and minority faculty.* New York: SUNY Press.

Downton, J. V. (1973). *Rebel leadership: Commitment and charisma in the revolutionary process.* New York: Free Press.

Goode, J. (2004, March 30). The mommy candidate. *Chronicle of Higher Education Online.* Retrieved September 17, 2007, from http://chronicle.com/jobs/2004/03/2004033001c.htm

Guldner, G. T. (2006). *Long distance relationships: The complete guide.* Corona, CA: J. F. Milne.

Hersey, P. (1985). *The situational leader.* New York: Warner.

Kotter, J. P. (1990). *A force for change: How leadership differs from management.* New York: Free Press.

Mannix, M. (2001). Take me, take my spouse. *ASEE Prism Online, 11(3).* Retrieved December 1, 2007, from http://www.prism-magazine.org/nov01/takeme.cfm

Villalba, C. (1999a). Partners in science, Part One: Negotiating the academic job market as postdocs. *Science.* Retrieved December 1, 2007, from http://sciencecareers.sciencemag.org/career_development/previous_issues/articles/0000/partners_in_science_part_one_negotiating_the_academic_job_market_as_postdocs/

Villalba, C. (1999b). Partners in science, Part Two: Looking for and negotiating faculty jobs together. *Science.* Retrieved December 1, 2007, from http://sciencecareers.sciencemag.org/career_development/previous_issues/articles/0000/partners_in_science_part_two_looking_for_and_negotiating_faculty_jobs_together/

von Ruschkowski, E. (2003). The 700-km commute. *Science.* Retrieved December 1, 2007, from http://sciencecareers.sciencemag.org/career_development/previous_issues/articles/2240/the_700_km_commute/

Wilson, R. (1999). The frustrating career of the 'trailing spouse.' *Science.* Retrieved December 1, 2007, from http://sciencecareers.sciencemag.org/career_development/previous_issues/articles/0210/reprints_the_frustrating_career_of_the_trailing_spouse/

Part IV
Flowering and Seed Forming: Leadership Models

Chapter 9
Emerging Leadership Through Experiences in Unfamiliar Academic Waters: Researching Invisible Truths

Claudette Giscombe

In matters of style, swim with the current; in matters of principle, stand like a rock.
~ Thomas Jefferson (Quotation by Author, n.d.)

Researching as Leadership Activity

For me, leadership involves having the wisdom to know when to swim with the current, namely a social context. The wisdom also involves knowing when to stand like a rock, which could entail taking different risks as a leader to research invisible truths. Leadership is a term that evokes many different meanings for different people. A literature quick search on the term leadership delivers thousands of definitions and an array of approaches to leadership. For centuries, we have asked the questions: Am I born a leader? Does leadership evolve? Is leadership associated with positional power? Can anyone anywhere be a leader? I adopt a more evolving view of leadership to frame my story – Leadership as an emerging researcher self process.

Mead (1934) attributed the development of one's identity to relation with the social experience, and as experience progresses, the self emerges.

> The self is something which has a development; it is not initially there, at birth, but arises in the process of social experience and activity, that is, develops in the given individual as a result of his relations to that process as a whole and to other individuals within that process. (p. 135)

As my experience progresses, the story of my emergent researcher self helps develop my leadership perspective. Additionally, one of Mead's ideas especially relevant to my story is that of an increase in consciousness, immediate consciousness, and self. Mead (1934) provided a foundation for understanding emergent, complex, response-driven processes like researching.

C. Giscombe (✉)
GrowthPerformanceAchievement-Academic Solutions Inc.; Formerly University of Massachusetts Amherst, Amherst, MA, USA
e-mail: cgiscombe@aol.com

K.C. Wieseman, M.H. Weinburgh (eds.), *Women's Experiences in Leadership in K-16 Science Education Communities, Becoming and Being*, DOI 10.1007/978-90-481-2239-4_9,

I make a case for a perspective on leadership that researching is an ultimate context of leadership that has power in generation of knowledge and the potential to lead self and others (Tichy, 1997). Research by definition creates new knowledge; knowledge is power; and, "leadership involves the exercise of power.. . .[and] power is the ability of one person to influence the behavior of another" (Bone & Kurtz, 1993, p. 230). This relationship between research, knowledge, power, and leadership is validated by the essential functions and skills an effective leader must acquire to retain and use the inherent powers associated with the status of a leader (Bone & Kurtz, 1993; Davis, Skube, Hellervik, Genelein, & Sheard, 1996). For example, an effective leader in research must be able to provide direction, influence and motivate others, and foster team work (Carey, Watson, & Carey, 2007; Tichy, 1997). Moreover, effective leaders in research show that they can lead courageously, champion change, coach and develop team members (Carey et al., 2007; Tichy, 1997). All these attributes are manifestations of the effective application of power. My experience as a researcher is leading me to conclude that leadership is essentially the process of motivating and influencing others to think in particular ways and/or perform activities designed to achieve specific objectives. This is also in essence a part of the researcher's role. This perspective is important because it frames my thinking about the process of my emerging leadership. At first I was hesitant to identify myself as a leader because I did not originally identify the research context as demonstration of leadership. Retrospectively, my experiences in research which have allowed me to believe in my innate abilities, define self, take initiatives, and stand on matters of principle, have helped me to develop my emerging leadership perspective. As a result, the knowledge attained from this research is a product of this leadership.

I think that leadership can take form as a person or persons moving ideas forward to change paradigms. It is not just a goal-oriented activity to move a project forward. Nor is enactment of a concrete entity based on visions and goals. Research can also be a person/persons moving ideas forward in particular, new and/or different ways. The research in which I specifically am involved focuses on shifting paradigms, and ultimately can influence and reshape projects, concrete entities, societal activities, and events. I am processing paradigm shifts about the connection between research and leadership. This shifting is what I am making public and known to an outsider. I think that leadership starts with ideas and this aspect is typically a private mental activity. This mental activity may not be the focus of most people's immediate conceptions of leadership. In my chapter I make visible this mental activity which is centered on idea construction and the shaping of the new ideas, as well a deep caring of how my ideas are used to address issues related to minorities in science education. I expose some of my ideas to public view, so others can do something with them. Obviously I hope that the ideas/new knowledge will be used in positive ways. This activity involves risk taking, which the leadership literature writes about as a leadership attribute.

In this light, I offer the following ten guidelines as encouragement for researchers to become leaders in spite of the tensions associated with researching.

1. Believe in Yourself: Believe in your innate abilities to persevere, navigate, and negotiate unfamiliar academic waters and chart the unknown in the research world.
2. Embrace the Right Attitude: Have an appetite to learn and to be a lifelong learner.
3. Be Prepared: Be ready to handle the unexpected jolts and disconnects in social context in relation to personal beliefs.
4. Take the Initiative: Have personal aspirations and seize opportunities to question and understand what is truth versus facts (Clark, 1989).
5. Stand Like a Rock: In matters of principle do not be afraid to ask the hard/sensitive questions and pursue solutions.
6. Be Courageous: Have the audacity to deviate from established ways of thinking.
7. Value Your Thoughts/Ideas: Put weight on your perspective, even prior to or without collegial recognition, respect, and status in academic circles.
8. Define Self: Know yourself, your passions, and your survival needs and balance that with your commitments to personal, moral, and social responsibility.
9. Take Risks: Be honest with yourself about the possible risks, but do not be afraid to test the waters and rely on the lenses of your critical friends for objectivity and insights.
10. Take Action: Value the process; each step of leadership in research has worth.

These guidelines for an emerging leader are highlighted in this chapter as I discuss my experiences, reflections, and response to my journey in research. As I reflected with my critical friends, family, and colleagues about each step of my journey and my hesitation to consider myself a leader, I have concluded that my journey represents a gradual step in leadership and has worth. The merits of leadership should not only be measured at the final destination but by the courage demonstrated in beginning the journey, and as each significant and/or important step is completed.

My Story as Theory Building

In 1973 two female islanders, a friend and I, arrived in the United States as immigrants who had already successfully overcome the first hurdle of emigrating from Third World countries. Our friendship and drive grew because we shared many cultural, social, and religious ties. As survivors of college experiences in the rigorous preparation of an undergraduate degree in the sciences during the post-civil rights era, we often reflected on their unfamiliar academic experiences of cultural, racial, and gender isolations in their science courses. Not fitting the majority, traditional White-male characteristics seen in most physics classrooms, we experienced loneliness, repeated racial and gender isolation, and unfair academic practices. We sat shocked as our biochemistry instructor described his sincere beliefs (stereotyping) that some groups were not capable of keeping up with the challenges/rigors of the course. He then iterated that students with limited preparation and background, like us, should take solace in the fact that we would not be the first to dropout.

Thrown into this new environment we often felt like we were drowning in an unfair academic body of water. We confessed that we, by virtue of our prior academic accomplishments and arrival to pursue science careers in reputable universities in the United States, thought that we were excellent academic swimmers. However, we soon realized that the waters in this new arena presented unfamiliar issues and challenges for which we needed new personal and pragmatic survival skills. However, we were not deterred from taking the required courses labeled hard sciences such as physics, as opposed to those labeled softer, gentler, more female science classes, such as biology. With the academic obstacles encountered we marveled at our tenacity, resilience, buoyancy, and the defiant strategies we both used to navigate and negotiate pathways to successful science careers. We attributed our successes primarily to: believing in ourselves and our innate abilities to persevere, navigate, and negotiate, and be tenacious and resilient; having an appetite to learn and be lifelong learners; being prepared for unexpected jolts and disconnects; and having the courage to deviate from established ways of thinking about stereotypic roles and career norms for Black women. Our accomplishments we believed: (1) added to the voices of African American women representation in science; (2) provided a counter-narrative (Delpit, 2003) to societal perceptions and theories about the inability of Black women to achieve in science and in science careers; and (3) allowed us to develop skills needed to function in different and vast academic waters. Ironically, our scientific experiences, which allowed us to develop leadership qualities such as defining self and developing the courage to ask hard/sensitive questions and to seek solutions, drove us into different career paths – medicine and science education, but our purposes remained similar. My friend dedicated her service to educating minority women about medical issues that impact them directly, while I have devoted my life career to educating minorities with a focus on Black women in science and saving minds.

As I began my lifelong journey and career, my role in the classroom changed from being a student to being a teacher in science. It was evident to me that my undergraduate experiences were not only transformative, but they began to shape my life, my passion, and career path. I believe that it was as a result of my undergraduate experiences in science (as described above) and my personal aspirations and achievements that I developed an insatiable desire to seize every opportunity to understand the truth about equity, access, and opportunities to quality science education. The statement that follows, which was adopted from Dewey's (1944) encompassing philosophy of education, summarizes my philosophy of education. "Education serves a social function: it is a social elevator from economic quagmire, and provides a plethora of opportunities for personal and societal good" (Giscombe, 2007, p. 4).

What was the truth that I desired to find? I believe it was the knowledge of a social context, an obscured reality that Clark (1989), Kozol (1992), and Morris (2004) shared about the spirit, educational experiences, abilities, resilience, and resistance of a significant group of minority achievers in education and in science.

This truth, that is not necessarily visible, is important and valuable in understanding tacit relationships between under-representation of minority groups in science and lived experiences. Scholars, like Rodriguez (2001), suggested that the present preoccupation and obsession of research that focus on achievement scores, and the national statistics (National Science Foundation, 2004) of the glaring gaps of under-representation of minority groups such as African Americans in the natural sciences, can only provide partial visible truths about the significant factors that shape the phenomenon of glaring gaps in achievement scores.

My opportunity to search for additional truths about this phenomenon came when I was a doctoral student. I was asked to join a team of experts comprised of researchers, scientists, science teachers, and middle school science students who received funding for a grades K-12 initiative to reform science teaching in public schools. As a graduate student serving as project assistant, teaching assistant, graduate student researcher, and the only African American on the project, I was intimately involved in many facets of the project. I was effervescing and exuberant about the endless possibilities and opportunities from this collaborative effort of researchers, scientists, science teachers, and students. In my development as a researcher, I had an appetite for being a lifelong learner. I embraced the right attitude and was confident that I had finally gotten the opportunity to be a learner and contributor on a team of experts that would make a difference in addressing issues around the under-representation of minority groups in science. I had had a personal interest in understanding these issues since my undergraduate science education experiences. Working with this project was the opportunity of a lifetime, a dream come true. You can imagine how I wanted to make every moment count as our rich and diverse team of experts explored real solutions to the daunting problem of under-achievement of minorities in the sciences.

Our basic approach to the problem was to design and model a unit on human and environmental issues, using inquiry-based instruction (Krajick, Czerniak, & Berger, 1999). The goal was to help teachers to see the relevance and importance of quality classroom science experiences that made practical connections between environmental issues in the local community to broader national and global issues. This unit to be used in teachers' classrooms would be evaluated for its effectiveness and value in complementing classroom practices in science. Another facet of the project in which I was involved was researching. The goal was to use our research findings about students' beliefs about how the approach affected their attitude and motivation to learn science to assist science teachers in planning and delivering middle school science curriculum.

We devised a unit that required collaboration between all the scientists and teachers in exploring local environmental issues. The community selected for study was close to the site in which our inquiry-based course was taught. The schools within this community were characterized as urban, where the majority of the students were African American and the majority of the teachers were White females. Nine of the eleven participating teachers were White females; four of these nine White female teachers were from the urban community selected for study. The interactions and outcomes based on these demographics led to unexpected jolts.

We began the unit with the science teachers and scientists going on a fact finding/exploratory mission about the community and local school district. Required activities included researching; collecting photos/sketches; visiting the community's local park, ponds, streams, vegetation; and, sketching a map of the location and its resources.

The teachers were non-resistant to all required activities except the one activity that required a one-day visit to the community's park. Unexpectedly, even the teachers teaching in this urban school district expressed fears/concerns about going to the park located less than ten minutes from the school where we held our weekly course meetings. The issue articulated was that the women feared for their own safety in the unfamiliar neighborhood within which a number of them earned their living. Although I believed that their perceptions seemed real and genuine, I was curious to know if they were based on factual evidences or stereotypes. I was even more shocked when one of the women agreed to participate only after having received permission from her husband. She stated her conditions, a one-time venture and an early departure. My experience as a leader/researcher was tested by the unexpected jolts that the reactions to the assignment produced. Blatant differences in attitudes and school science activities led me to conclude that science teaching and learning experiences and by extension the learning outcomes in schools were undeniably different. This indeed was my first wake up call.

As a teaching assistant, I knew that the community activity had been carefully researched and planned. The team on which I participated had done all the preliminary planning for this activity. During our planning and preparation for the activity the team did not voice any concerns about safety. From having taught in major urban cities in the United States, my post-assessment of the teachers' concerns led me to conclude that there were no significant evidences of safety issues for this park or its location in the local community. Because of my position as a graduate student and because of the sensitivity of the issues, I was unable to further explore the fears articulated by most of the female teachers. Through informal conversations and observations I came to believe that the teachers were afraid of an unfamiliar environment with which they were not culturally or emotionally connected. This discomfort and unfamiliarity resulted in a disconnect from the need to achieve a desired goal in the learning process – students making meaningful educational connections.

I was faced with a dilemma which tested every fiber of which I was made. Metaphorically the dilemma was, do I swim with the current or stand like a rock? What was my professional, moral, and social responsibility? Did I have the courage to deviate from the established ways of thinking and swim in unfamiliar academic waters, thus charting the unknown? Using my ten guidelines as the basis of personal reflection has helped me determine my emerging leadership in the research context. Originally, like all the other researchers on the team, I was interested in understanding the phenomenon of the science gap from a traditional perspective, i.e., from the theoretical underpinnings of: (a) the need for change in classroom pedagogy, (b) increased inquiry-based learning, and (c) effective intervention programs. But the unexpected jolt from personally experiencing a culturally and racially related reality disconnect between African American urban students and their White female teachers, seen in the lack of meaningful partnerships and connections between the teachers and their students and the inequities in access to quality science learning environments, forced me into being courageous, standing like a rock, having the audacity to deviate from the established ways of thinking. I began to see the need for additional data and analysis that aligned with individuals' experiences in science education. Thus began my emerging self process in leadership through my research endeavors.

Instead of finding solutions, I began asking hard sensitive questions about the value of research methods that are limited in their abilities to provide contextually rich data and analysis of access and opportunities of minority groups to quality meaningful science experiences. I concluded that the team of experts' research approach could not provide explanations about the ways in which socio-cultural disconnects of "academic expectations; demands, approval, and or condemnation of others" (Dewey, 1944, p. 12) impacted the attitudes, beliefs, and learning outcomes in science education. As I looked more closely at the failing scores – gap gazing (Rodriguez, 2001) – and solutions for issues regarding national achievement scores, I became even more frustrated. Although I knew these scores were factual and represented a specific set of data that provided important knowledge about achievements in science, I was left with unanswered questions. What do the lived experiences of minorities in science classrooms look like? Why is the gap widening in spite of the myriads of intervention models and mentoring programs resulting from research in science reform?

I started to think that in addition to the findings and analysis of achievement scores, the statistical facts around race and gender, and the significant differences in representation of these groups in the natural sciences, there was an urgent need for a broader perspective, new ways of theorizing, and a paradigm shift in our science reform efforts. I wrestled with finding invisible social truths and "subjugated knowledge" that are "governed by institutionalized forces that control and frame" (Foucault, 1982, p. 213) the discourse. For my leadership development it was important to question and understand more than just the statistical facts around the widening achievement of the gap between dominant and minority groups. Maybe in the current research obsession of focusing on learning outcomes and the underrepresentation of minority groups in science, there was an underlying assumption that the most significant student learning determinants were linked to instructional practices.

Ultimately I knew I wanted to make a difference in research but my fear of deviating from the established ways of thinking about a phenomenon as well as the daunting fear of deviating from experts' assumptions and perspectives almost crippled my personal passion. Also, there was a temptation for mediocrity and a self-serving interest to only satisfy my doctoral requirements. This decision would result in taking the line of least resistance in researching, and defending a simple, mainstream, non-controversial, academic dissertation topic, rather than trying to solve the problems found in the academic world. Another of my concerns was risking not being heard in the arena of scholarly research. I questioned, will anyone put weight on my perspective which represents a paradigm shift and a new way of thinking, not developed by known academic scholars? This inner turmoil was compounded by my limited financial resources, and weighed heavily on my decision to take the risk of being alone in researching. As my mental and moral struggles intensified and through much rationalization I consoled myself with the idea that this was not the time in my career to make waves. First, I needed to gain collegial recognition and respect from the *Who's Who* in academia circles, and then I could venture into unchartered academic waters and be better positioned for my voice to be heard. Such reasoning did not last before I was confronted with being honest with myself about my personal responsibility and my commitment to moral and social responsibilities. I pondered; can truths be adjusted to make palatable the reality of what I was experiencing? This mental activity, paradigm shifting to address issues related to minorities in science education involves risk taking, which a leadership attribute (Bone & Kurtz, 1993; Tichy, 1997).

In the midst of this dilemma I was reminded of my undergraduate experiences in the sciences. Only now I was the science educator confronted with a set of circumstances which would test my leadership and my moral compass. I began to feel that the metaphor of drowning in unfamiliar water conditions, which I used earlier to describe my undergraduate science experiences, again best described the unexpected experience I was encountering on the urban initiative national project in science reform.

Using this imagery or metaphor, I likened the team of experts in the funded grant project to professional rescue teams who were called at a critical point in our nation's desperate effort to find solutions to avert the tragedy of loosing potential scientists from the science academic waters. In this context academic waters represent that in which we are immersed and that from which our actions come. We operate from that which we know and that becomes who we are. Instructional life-lines for academic rescue represent the frantic trials by science educators to reach students through new and innovative classroom practices. Finally, shore and land are used interchangeably to represent the community of successful students in science education.

At the tragic scene was a group of intelligent student swimmers, destined to drown, who needed to be pulled onto shore by the team of science experts. Once on shore the rescued students would join other groups who had already successfully negotiated and navigated deep, unfamiliar academic waters to safety and joined the ranks of science achievers on land. It should be noted that earlier heroic rescue efforts of just science teachers, first present at the tragic scene, were unsuccessful. Successful rescue required the collaboration and expertise of professional scientist, researcher, and science teacher rescue teams. It was for this reason that the team of experts in our project was called to the scene. We were experienced academic life savers with diverse skill sets, expertise, and training in student swimming abilities who could provide effective lifelines. We were trained to understand, or maybe we believed that we understood all the environmental conditions and climatic issues. Our team was commissioned to exhaust every possible strategy and to find ways in which we could rescue each student in the water. After much debate and analysis of similar and previous rescue efforts, students' condition, and the environmental factors at the tragic scene of the urban middle school district, our team proposed strategies we had repeatedly used and deemed best for this target group and situation.

Unfortunately, after much rescue effort, our mission also failed. Although most of the team experts were skilled professionals and academic swimmers; understood how to use needed lifelines; and had navigated unfamiliar academic waters before, we had never experienced training and navigation for this type of water. We had never encountered the murkiness of the historic existing social differences in these waters.

From my assessment of the failed mission, in the context of the community activity of the project, it was evident that the students were drowning in academic waters unfamiliar to educators, science educators, and scientists. What knowledge did we need that was significant for the rescue mission in this project? We thought the students were drowning because they could not reach the instructional lifelines. Ironically, the lifelines were close by and within reach, but the real problem was that the invisible yet significant murkiness of the waters prevented students' forward movements toward the lifelines. Our team did not have the skill sets to execute an effective rescue effort because they did not recognize these unique aspects of the academic waters in which the students were immersed and trapped.

I too focused on instructional lifelines and was puzzled by the fact that the lifelines, which were at arms length and within the practical reach of each swimmer, were not being reached. My obsession with the conscious, visible classroom conditions; the effective pedagogical approaches; and role models and mentors, obscured some of the underlying conditions that served to facilitate or deny access and opportunity in the science navigation experience. Like me, most of the team of experts had only been scratching the surface for answers and were missing some of the significant success determinants in the underlying academic conditions. The condition which was most significant to impeding the science rescue process was not visible and tacit relationships were sometimes missed.

In retrospect, my experience with the teachers' strong emotional discomfort in visiting the local park undeniably nudged me into thinking about other tacit, yet significant truths that determine the success of minority groups negotiating and navigating in science education. I wondered if providing effective instructional lifelines for students stranded in the waters of science failure would only work if the students were able to overcome the difficulties of social ills in the unfamiliar waters in which they were trapped. Undoubtedly in my mind, my leadership emerged as a result of this experience, my moral strength, and my unwillingness to adjust the truth about social-cultural disconnects – "academic expectations; demands, approval, and or condemnation of others," (Dewey, 1944, p. 12) and the impact on learning science. The unconscious influence of disconnects in this project which were so subtle yet pervasive can affect attitude and learning in science education (Dewey, 1944; Morris, 2004).

The mental turmoil I was experiencing subsided as I developed the necessary courage and trust in my ability to make a difference. I began to value my ideas, putting more weight on my own perspectives, even prior to or without collegial recognition, respect, and status in academic circles. My research perspective shifted from focusing on gap gazing (Rodriguez, 2001) and classroom strategies to focusing on a broader scope of knowledge generation which included the actual lived experiences of African Americans as they navigated to success in science education. I believed this shift would enhance the discourse and understanding of the similarities and differences in equity, access, and opportunities between white and non-white groups in the natural sciences. As an emerging leader, I was no longer afraid to leap into researching and charting the unknown; leave the familiar behind; and take risks to map my career and continued emergent leadership.

As I reflected on these experiences, my increased inner tension and professional turmoil, I believe that in response I developed the characteristics of courage, honesty, and risk taking which have been cited as attributes of leaders (Carey et al., 2007; Tichy, 1997). These characteristics ultimately framed my course of action and resolution to the problem of using a different set of lenses to view the phenomenon of the under-representation of African Americans in the natural sciences. My course of action for my research agenda emerged from: (a) developing the courage to deviate from the current trends in research in science reform; (b) overcoming my fears of having ideological differences from mainstream ways of theorizing, which helped me develop the courage to deviate from the team's perspective and research focus on instructional classroom practices; (c) being honest with myself about my passion, and moral and social responsibility to make a difference in inclusive practices in science; and (d) resolving to take an academic risk and a leap in spite of the possibility of negative consequences in my career advancement and in my financial base.

I began by looking for research studies and scholarly work that would provide support for the research direction I wanted to pursue. The works of scholars like Denzin and Lincoln (2005), Glesne (2006), Lightfoot (1983), Lincoln and Guba (1985), Scheurich, (1998), Schutz (1967), Seidman (1991), Smith Tuhiwai (1999), and van Manen (1990) and their ideologies related to subjective understanding not only gave me support but also provided a foundation from which I could argue that: (a) using research methods that focus on aggregate numbers to explain all social phenomenon are limited in their abilities to provide contextual rich data and analysis of differences in social experiences, like that which I had encountered on this project and (b) there was an urgent need for using a different set of lenses to address the issues around the phenomenon of the lived experiences of African Americans, particularly African American women in the natural sciences.

The next step in my plan of action was to build on the existing works of scholars such as Hine (2005) and Warren (1999) that shed light on the experiences of Black women scientists, particularly in terms of their struggles, strategies, successes, and scientific contributions. As I began exploring the social, historic, educational, and scientific literature, I was unable to find research which focused on the contextual realities of historically marginalized and disenfranchised African American women in the sciences. This limitation led me to my next strategy in which I drew from the findings of related works such as Delpit (2003) and Morris (2004) for my theoretical underpinnings. Although, the lack of literature to support my assumptions created a temporary setback, I forged on with confidence, continuing my testing of unfamiliar academic waters.

My next goal was to balance the objectives and expectations of the team with my own passion to pursue research that explored the experiences of African Americans in the sciences. As a member of the team, I continued to meet my prior research obligations. In addition, I found literature that focused on the social interactions, attitudes, and meanings students make of their academic/science experiences. I explored the lived experiences of one of the participating urban middle school students to determine her beliefs about her journey in science education. This student, an African American female, was selected from an in-class focus group because her story was of particular interest. She was very vocal about contrasting her past unfulfilling science experiences with present positive science experiences. She stated, "Usually I would not like science... but since this year it is more hands on... field trips, it's a lot of fun." She continued to describe herself in this science class as "very highly educated... I use very high scientific vocabulary... walking in the shoes of scientists." She spoke about the fun she had when searching for chemicals in the playground in the park. She believed her field trip in the park increased her understanding and confidence in science. She said:

> Yes, I remember when I went home (from the field trip) I discussed this over with my parents and my parents were like highly interested. They were worried when I said arsenic because we have a deck made out of pressure treated wood and we checked it and it had arsenic in it.... My mother didn't really understand like some of the vocabulary words and I had to explain to her.... Like sample was as an example.... But, I said its different, a sample you have, actually holding on pieces... an example is... just written down or it could be standing or described. (C. L. Giscombe, personal communication, April 30, 2003)

I presented the preliminary findings from this case study as a mini pilot investigation to a jury of my peers in two of my doctoral courses. From their feedback I would determine if my argument, that research methods which only focus on aggregate numbers to explain all social phenomenon are limited in their abilities to provide contextual rich data and analysis of differences in social experiences. If so, the argument would be significant enough for further investigation in my dissertation research.

My peers' responses engendered such enthusiasm and discussion that I felt like I was finally beginning to explore the truths for which I was searching, truths around the phenomenon of the underrepresentation of African American females in science. With increasing confidence I continued to test the uncharted by presenting the initial findings from my case study in presentations at science education conferences. The enthusiasm from the collegial discourse in these forums helped me to believe in myself. It gave me confidence in the value of this new way of theorizing about the underrepresentation of African Americans, in science, particularly women.

The enthusiastic collegial responses also gave me the boost of energy needed to seek another form of academic endorsement, that of recognition through academic scholarship. My goal was twofold: first, I wanted to compete with others for grants and scholarships designed to assist individual researchers who have demonstrated leadership in research; and second, I wanted to get financial support for my research. I was successful in both endeavors and received substantial funding in support of my dissertation from an established endowment. Although the researching path was lonely and unpredictable, the knowledge I was generating was now deemed valuable in the discourse in science reform.

This risky but productive journey led me to craft the research design for another significant yet not final step in my leadership and professional journey – the preparation, completion, and defense of my dissertation. In my dissertation I explored the life experiences, personal stories, and reflections of African American women science faculty in higher education. My objective was to understand the contextual realities within which the opportunity for their legitimate participation in the natural sciences has been facilitated or denied. I sought to gain insights on how the participants believed that they had negotiated pathways to achieve and contribute to this point in their science careers. This study provided insights into understanding the connections between life experiences, educational processes, social realities, and the meanings African American women attach to their accomplishments.

Leadership Response

My pilot study, namely the case study of one urban middle school female student who was African American, was transformative. Researching heightened my courage, increased my confidence, supplied the needed energy, and gave weight to my perspectives as an emergent leader. Researching allowed me to take greater steps and more risks in generating discourse on a topic even prior to or without collegial recognition, respect, and status in academic circles. The welcomed and unexpected interest and positive responses from diverse groups of professionals at national conferences, science educators, members of African American churches, and leaders in the African American communities has provided me the ultimate confirmation of the value of my work and of my emergent leadership.

The 10 guidelines mentioned at the beginning of the chapter point to the emergence of a researcher's leadership. Sashkin and Sashkin (2003) reported that leaders who proactively empower others by providing the knowledge, resources, and access to get the job done are effective leaders themselves. I suggest that researchers who acquire the knowledge, resources, and skills to commence innovative research are demonstrating growth in developing their leadership skills (Antonakis, Cianciolo, & Sternberg, 2004; Clinton, 1988, 1998; Northouse, 2007).

The body of knowledge in which I have immersed myself was generated from the voices of the African American women scientists. Researching the lives and experiences of African American women scientists is an act of leadership that is resulting in motivating and influencing others to engage in different ways of thinking about the issues related to minority achievements in the natural sciences. This leadership and response driven process has the potential to add to the body of knowledge for scholars and other leaders in academic fields such as education, science education, natural sciences, sociology, and anthropology. It is therefore my strong belief that through continued leadership and collaborative efforts, a strong, invaluable contribution can be made toward the science reform efforts.

References

Antonakis, J., Cianciolo, A. T., & Sternberg, R. J. (Eds.). (2004). *The nature of leadership*. Thousand Oaks, CA: Sage Publications.

Bone, L. E., & Kurtz, D. L. (1993). *Contemporary business* (7th ed.). Fort Worth, TX: Dryden.

Carey, P., Watson, E. D., & Carey, P. A. (2007). *Plain talk about leadership: How to develop the values, knowledge, skills, attitudes and perspectives for effective leadership*. Charleston, S.C.: BookSurge Publishing.

Clark, K. B. (1989). *Dark ghetto: Dilemmas of social power*. Lebanon, NH: University Press of New England.

Clinton, J. R. (1988). *The making of a leader: Recognizing the lessons and stages of leadership development*. Colorado Springs, CO: NavPress.

Clinton, J. R. (1989). *Leadership emergence theory: A self-study manual for analyzing the development of a Christian*. Altadena, CA: Barnabas.

Davis, B. L., Skube, C. J., Hellervik, L. W., Genelein, S. H., & Sheard, J. L. (1996). *Successful manager's handbook: Development suggestions for today's managers*. Minneapolis: Personnel Decisions International.

Delpit, L. (2003). Educators as "seed people" growing a new future. *Educational Researcher, 7*(32), 14–21.

Denzin, N. K., & Lincoln, Y. S. (Eds.). (2005). *The Sage handbook of qualitative research*. Thousand Oaks, CA: Sage Publications.

Dewey, J. (1944). *Democracy and education: An introduction to the philosophy of education*. New York: The Free Press.

Foucault, M. (1982). Afterword: The subject and power. In H. L. Dreyfus & P. Rabinow (Eds.). *Michel Foucault: Beyond Structuralism and Hermeneutics* (pp. 208–226). Chicago: University of Chicago Press.

Giscombe, C. L. (2007). Pathways to success in science: A phenomenological study examining the life experiences of African American women in higher education (Doctoral dissertation, University of Massachusetts, Amherst, 2007). *Dissertation Abstracts International, 68*, 4660.

Glesne, C. (2006). *Becoming qualitative researchers: An introduction (3rd ed.)*. Upper Saddle River, NJ: Prentice Hall.

Hine, D. C. (Ed.). (2005). *Black women in America*. Oxford: Oxford University Press.

Kozol, K. (1992). *Savage inequalities: Children in America's schools*. New York: Harper Perennial.

Krajick, J. S., Czerniak, C. M., & Berger, C. (1999). *Teaching children science: A project-based approach*. Boston: McGraw-Hill Press.

Lightfoot, S. L. (1983). *The good high school: Portraits of character and culture*. New York: Basics Books.

Lincoln, Y. S., & Guba, E. G. (1985). *Naturalistic inquiry*. Beverly Hills, CA: Sage Publications.

Mead, G. H. (1934). *Mind, self, and society*. Chicago: University of Chicago.

Morris, J. E. (2004). Can anything good come out of Nazareth? Race class and African American schooling and community in the Urban South and Midwest. *American Educational Research Journal, 41*(1), 69–112.

National Science Foundation, Division of Science Resources Statistics. (2004). *Women, Minorities, and Persons with Disabilities in Science and Engineering*: 2004. NSF 04-317 (Arlington, VA, 2004).

Northouse, P. G. (2007). *Leadership: Theory and practice* (4th ed.). Thousand Oaks, CA: Sage Publications.

Quotation by Author. (n.d.). Retrieved September 13, 2008, from http://www.quotationspage.com/quotes/Thomas_Jefferson/

Rodriguez, A. J. (2001). From gap gazing to promising cases: Moving toward equity in urban education reform. *Journal of Research in Science Teaching, 38*(10), 1115–1129.

Sashkin, M. & Sashkin, M. G. (2003). *Leadership that matters: The critical factors for making a difference in people's lives and organization's success*. San Francisco: Berrett-Koehler.

Scheurich, J. J. (1998). Highly successful and loving, public elementary schools populated mainly by low-ses children of color: Core beliefs and cultural characteristics. *Urban Education, 33*(4), 451–491.

This reference is not cited in the text part. Please provide.

Schutz, A. (1967). *The phenomenology of the social world* (G. Walsh & F. Lenhert, Trans.). Chicago: Northwestern University Press.

Seidman, I. E. (1991). *Interviewing as qualitative research: A guide for researchers in education and the social sciences*. New York: Teachers College Press.

Smith Tuhiwai, L. (1999). *Decolonizing methodologies: Research and indigenous peoples*. London: Zed Books Ltd.

Tichy, N. M. (1997). *The leadership engine: How winning companies build leaders at every level* (1st ed.). New York: HarperCollins Publishers.

van Manen, M. (1990). *Researching lived experience: Human science for an action sensitive pedagogy*. New York: SUNY Press.

Warren, W. (1999). *Black women scientists in the United States*. Bloomington: Indiana University Press.

Chapter 10
Teaching Elementary School Science: The Road Less Traveled

Kathleen S. Davis

The Story of Becoming a Leader–Mentor of Elementary Science Preservice Teachers

I often think of my leadership role as a kind of mentoring. When describing mentoring, Browne (no date) contends:

> Many people in academics confuse mentoring with being an academic advisor. Being a mentor is much more than talking about what classes a student needs to take to meet the requirements for a degree program. Mentoring is [taking] a stand for the success of a student, for the accomplishment of their dreams and the generation of their powerful futures. It is believing that they are capable of succeeding; it is assisting them, bridging gaps when the gaps appear insurmountable. It is listening to them, being their guide, their friend, their advocate and a catalyst for their development. (p. 1)

So leadership is not so much about dragging individuals along, but it is more like putting them on roller skates and giving them an open pathway to travel in order to accomplish their goals.

I have found that in order to enable elementary preservice teachers to become teachers of science more is required than the standard science methods course. What is needed is a form of mentoring that not only illuminates science as inquiry and connected knowing, but also, provides preservice teachers with sponsorship, challenges, coaching, and a look into their possible selves (Packard & Nguyen, 2003). Furthermore, to mentor, as a form of leadership, requires one to know the challenges that new science teachers face (Davis, Petish, & Smithey, 2006) and to acknowledge the socio-cultural beliefs and practices associated with science and schools and the hierarchies and hegemonies that exist in these settings. Such mentoring fosters ways to enable newcomers to become legitimate members of those communities, creates change in those communities, and assists novices in becoming change agents.

My pathway, to this view of mentoring as leadership, began as an undergraduate in an elementary education teacher preparation program and developed as I taught

K.S. Davis (✉)
School of Education, University of Massachusetts Amherst, Amherst, MA, USA
e-mail: kdavis@educ.umass.edu

K.C. Wieseman, M.H. Weinburgh (eds.), *Women's Experiences in Leadership in K-16 Science Education Communities, Becoming and Being*, DOI 10.1007/978-90-481-2239-4_10,

science in elementary schools for 18 years, beginning in 1972. It further evolved as I made my way through graduate school and is now being tweaked even more, through reflective practice, as I teach preservice teachers at the university level.

I had always done well in my high school science classes, but did not see science as something connected to my everyday life or as a possible career. In 1970, I was a junior and an elementary education major at Eastern Illinois University. Though my academic concentrations were history and library science, I was enrolled in the required science methods class. Our professor was an energetic guy, always having us do something that perked our interest.

One day he asked us, "How deep, do you think, is the campus pond?" Most of us had a personal experience with the campus pond as, for one reason or another, students were frequently tossed into it by their dorm buddies. It was on the occasion of my wedding engagement that my friends from Lincoln Hall had picked me up, carried me out of the building and across the lawn, and heaved me over the rails of the wooden bridge. As I pondered the question, experience told me that the pond was dark, murky and as deep as Lake Michigan.

After some discussion, he said, "Well let's find out." As a class we constructed a plan of how we might map the depth of the campus pond – get a boat and, while rowing across the pond, drop a plumb line at regular intervals. Then, graph the results. During our planning time, I mentioned that I had never even been in a boat. The day of the activity, who do you think was placed in the boat first? What an experience I had, at the helm of science inquiry!

That semester, there were many such activities that invited me into the world of teaching elementary science – a walk in the rain to observe what was not as evident on a dry day; a springtime observation of the night sky; and many, open-ended experimentations with materials from the *Elementary Science Study* (ESS) and *Science Curriculum Improvement Study* (SCIS) kits of the 1960s. It was through these experiences that I was able to make a connection between the questioning and exploration that I had done as a child in my backyard and in my Girl Scouting to the kind of science teaching and learning that could take place in my classroom.

When I entered the field of teaching, I would not have self-identified as a science teacher. Yet, for my first position, I was hired to teach 5th–8th grade science in a country school in Illinois. Oh! The fun we had experimenting in my basement classroom, exploring the outdoors, taking field trips to ponds, parks, and observatories, and raising a litany of classroom animals! Over the next 18 years, I continued to do science with elementary students in Peoria and in a suburb of Chicago. I believed I had the best-kept secret – my job was fun and interesting; I was constantly learning with my students; and they found science to be fun and interesting, too.

Imagine my surprise when I learned that my colleagues wanted nothing to do with teaching science! "You go ahead and teach the science," they would say. They felt secure in their worlds of social studies, reading, and language arts. I found that, in general, many elementary teachers, the majority of whom were women, were not willing to teach science, and many students were not learning science as a result. Somehow they had not made the same connections with science I had made as an undergraduate and were missing out on an exciting aspect of teaching children. So, in 1989, I headed to graduate school to get an advanced degree that would allow me to teach preservice teachers full time and work towards turning this situation around.

However, at the time, I did not understand the complexities of institutional and societal sexism and gender-based oppression and how that related to elementary teachers and science teaching. In the fall of 1989, I had learned about a new funding opportunity from the American Association of University Women (AAUW) – the Eleanor Roosevelt Fellowship. With the application that came in the mail was a flier that was to change my life. It read:

> Your interest in math and science was sparked in elementary school, where books and tests used familiar examples and where you were called on a lot in class. In high school, you found role models and teachers who encouraged you to pursue your dream. You did well on college entrance exams and got into a good school. Now you're an architect, or a doctor, or an elected official, or a corporate executive. . .A Likely Story (AAUW, 1989).

This was not my story. In the flier, AAUW outlined how "girls are shunted away from the intellectual disciplines that might let them enter the key professions" (AAUW, 1989). They pointed to "faulty" teaching approaches, sexist practices, and "biased testing" as factors that led to "cheat[ing] girls out of their futures" (AAUW, 1989). This information shocked me! I knew nothing about these issues.

This began my critical examination of issues of gender in science and science education and my leadership in this area. Through my graduate studies, I began to better understand the hegemony of science and science education and the inequitable structures, policies, and practices of the culture. I better understood why I and other elementary school teachers came to teaching with a paucity of science content knowledge; why my colleagues did not want to teach science; and why my 5th grade female students would leave their dreams of science careers behind by 8th grade.

Since 1994, I have attempted to take the lessons learned from my teaching practice and my research studies and develop a way of mentoring as leadership with preservice and inservice elementary teachers. However, this work of scaffolding individuals into science activity is not without tensions and dilemmas. At times, as a mentor/leader, I feel like I am escorting individuals to their demise as one outlines the purported practices necessary for success in science and science teaching, all the while knowing that power structures and practices in schools and science will challenge preservice teachers' best efforts. Often, it can seem like I am encouraging individuals to "buy in" to the oppressive and dominant culture.

These tensions were made explicit in a play entitled "The Vampire Chronicles," that I viewed at AERA in 1993. Women dressed in academic robes with collars turned up in the image of the blood-drinking creatures of Anne Rice distinction walked onto the stage. The characters, and the participants in reality, were senior and junior faculty and graduate students. Stories outlined in novels – the vampire drinks the blood of its victim; the victim weakens and, out of necessity, becomes a vampire and continues the gory rites with new victims – were likened to the ways of academe – senior faculty bring newcomers into the academic community and use them to sustain the senior faculty member's research agenda. In turn, the graduate student and/or preservice teacher becomes a member of the community, only to repeat the abusive rites on others.

A vampire I try desperately not to be. To break this cycle requires mentoring and leadership. As Mary Lindsley, a member of the Women in Science support group I followed in my dissertation study, pointed out, it is important to offer support to others, to "not let people be *crushed* by the system," and to make the dynamics of the community explicit (Davis, 2001, p. 384). She stated:

> ...I might have ideas or insights that they might not have because they haven't been there....I can look at it from the other end, the power end of it...what could be going on...just give them that feeling about what's really going on....[S]ome of these situations that people find themselves in are one thing on the surface, but there's something else really going on....[J]ust having more experience and being older...understanding what drives some issues, but that just comes because...I've been at this longer and after that [it's] either because of what I've suffered through or what other people have suffered through (Davis, 2001, pp. 404–405).

Thus, mentoring as a form of leadership entails making explicit what is necessary for success as outlined by the dominant group. It also means illuminating inequitable structures, policies, and practices of the educational and science communities. It means making explicit the opportunities that exist, describing the possible strategies newcomers might employ to be successful in their careers and creating change where inequities exist (Davis, 2001).

Theoretical Framework

Several areas of the literature illustrate my pathway through science education and contribute to my leadership style and activities. One branch centers on access to science, a community of practice – and what that might entail for the female elementary preservice teacher. A second looks at the role mentors – supporters, institutional agents, and insiders to authentic science activity – can take to foster legitimate participation and what that means for me as a leader in elementary science teacher education.

Access to a Community of Practice

There are pathways and roadblocks to legitimate participation in a community of practice, such as science and science teaching (Lave & Wenger, 1991). Scholars (Bourdieu, 1986; Davis, 2001; Delamont, 1989; Harding, 1991) have illuminated many factors that contribute to the construction of boundaries within the science community and the subsequent insider/participant status for some groups and peripheral/outsider status for others. Community social structures, power relations, and requirements for legitimacy define what is possible for access and participation and serve as door openers or gatekeepers (Davis, 1999). For example, at the onset, individuals need some of the capital, such as science content knowledge, that is ultimately acquired through participation within a community in order to gain initial access and be considered legitimate. Secondly, biases enter at all stages of scientific inquiry; dominant groups usually define what counts as a scientific problem and most frequently determine the process and direction of investigations and research (Harding, 1991). Hierarchies result in newcomers having little decision-making power about their experiences, and the values, ways, and practices of the communities in which they wish to participate.

Mentoring

Educational institutions, businesses and industry, and professional organizations, and their members, and their connections to resources and credibility, can be critical to those who seek access to science and science teaching (Bourdieu, 1986; Delamont, 1989; Packard & Nguyen, 2003). Such social capital provides access to both learning opportunities and valued status and can be used to obtain jobs, funding, knowledge, access to powerful networks, and awards.

Mentoring is viewed as a common form of social capital. A mentor is usually perceived as an older person who is a veteran in the field – a role model, teacher, advocate, and counselor for novices (Casey & Shore in Packard & Nguyen, 2003). However, historically, mentoring has been exclusively reserved for men (Cullen & Luna, 1993). Such connections have been lacking for females in science and education (Delamont, 1989; Luke & Gore, 1992).

From a critical-feminist model, the dyadic, hierarchical positioning within the traditional mentoring relationship is seen as problematic for women (Packard, Walsh, & Seidenberg, 2004). Mentoring, where veterans clone newcomers into duplicates of themselves, is also seen as less than desirable. In contrast, such as in composite mentoring models, individuals of different backgrounds and experience (e.g., peers, colleagues, administrators) offer assistance, and co-mentoring and co-learning opportunities can be developed (Packard & Nguyen, 2003).

Mentoring can take several forms: sponsorship, challenge, and coaching – all of which can be critical to an individual's continuation in science (Packard, 2004–2005). In addition, similar to Browne's and Lindsley's beliefs about mentoring, individuals with positions of power within formal institutional settings can, as "institutional agents....provide or negotiate the provision of resources, support and opportunity for others" (Stanton-Salazar, Vasquez, & Mehan, 1995, p. 3) and provide forms of support so novices, like preservice teachers, might progress through institutional systems and have significant control over their lives and futures.

Important, too, is how mentoring relationships can foster women's goals and what they see as possible for their futures through validation and feedback. Individuals "can create, develop, or imagine possible selves through interactions with the significant adults in their environments...where [their] possible selves are continually developed and changed and the durability of desired possible selves relies heavily on the validation of significant adults..." (Packard & Nguyen, 2003, p. 261). Thus, providing acknowledgement and validation of preservice teachers' goals and dreams appears to be a critical element of positive mentoring.

Mentors as Change Agents

As a form of leadership, mentoring requires one to understand and acknowledge the socio-cultural beliefs and practices associated with science and schools and the concomitant hierarchies and hegemonies that exist in these settings (Davis, 2001). Oppressive educational and work settings can result in limited opportunities for preservice teachers to engage in the practices of schools and science and to acquire valued knowledge, skills, connections, and resources. So mentors need to make explicit the values, ways, structures, and practices of those communities and be willing to facilitate the acquisition of valued capital and support the activities and contributions of newcomers. Mentors need to legitimize other women's experiences; listen to their goals and needs; provide insight about bad or potentially bad situations; help novices take needed action; and outline ways to make change in their educational and professional settings.

In addition, mentors need to acknowledge their own struggles and those of their colleagues within the science and teaching community. Women science educators, working to facilitate females' participation in science activity, encounter many of the same roadblocks experienced by the females they attempt to support (Davis, 2002; Varanka-Martin, 1997). They also experience backlash within their communities as a result of their political activity. To be successful in this role, they will need to recognize bias and sexism within the science community, examine their own values, and determine what they are willing to risk in order to enact change (Brown, 1991).

It is important for mentors who work as change agents to establish structures and approaches that are inclusive, equitable, and participatory. Here females (1) can meet in collective groups and reflect on their needs and experiences as they relate to their community participation; (2) are able to voice what their experiences are within the science community, what structures, policies, and practices do or do not support their participation, what beliefs and practices conflict with their needs and values, and what should be changed; and (3) have decision-making power within the community (Young, 1990). Women, novices, and veterans, must be able to claim voice and power and challenge the inequities they encounter in their professional and educational settings.

A New View of Science Learning and Teaching

Framed in an inclusive pedagogy, science teaching and learning through inquiry provides a science learning community that is meaningful; views learners as science doers and knowledge producers; and supports the autonomy and decision-making power of all members (Davis & Falba, 2003; Roychoudhury, Tippins, & Nichols, 1995). Here, individuals have autonomy over the questions, problems, or phenomena to be explored and learn in authentic contexts using and integrating the knowledge, skills, and tools of science and other disciplines (Hurd, 2000). The science learning context is transparent, equitable, and participatory, honoring diverse voices experiences (Young, 1990).

To enable elementary preservice teachers to engage their students in science activity, novice teachers need contexts where they can reflect upon their preexisting knowledge and beliefs about teaching, learning, learners, and science (Borko & Putnam, 1996). It is important that they learn content and pedagogy through engagement in learning activity that mirrors and models the kind of experiences they would provide their students (National Research Council, 1996) and that they make strong links between personal learning and their classroom contexts (Anderson & Mitchener, 1994).

Leadership Activity – Designing an Invitation

I view my role as an elementary science teacher educator as providing preservice teachers with leadership through mentoring as they begin their journey of teaching science to children. The set of leadership activities described in this chapter serves as an invitation for all educators, but preservice teachers in particular, to reflect upon their beliefs about science teaching and learning. I use this set of activities in the methods course I teach, as an invitation to foster preservice teachers' engagement in science, their construction of new views of science teaching and activity and the development of personal identities of science doers and knowledge producers. These activities include:

- Preservice teachers reflecting about science and their science experiences
- Preservice teachers and the instructor drawing of a teacher teaching science to children
- A metaphor/simile written by preservice teachers describing what teaching elementary science is like and by the course instructor describing what mentoring is like
- Preservice teachers listing goals for the elementary science classroom and what a classroom meeting these goals would look and sound like
- The instructor listing goals for the methods course
- Preservice teachers engaging in guided and open inquiry using the 5 Es Learning Cycle (Atkin & Karplus, 1962; Bybee et al., 2006)
- Preservice teachers' teaching inquiry-based lessons
- Preservice teachers and instructors engaging in a community exploration and critical reflection

I have implemented this set of activities in my elementary science methods course and have found them to be successful in my leadership/mentoring of elementary preservice teachers.

Teacher Beliefs about Science and Science Teaching

At the beginning of the semester, most preservice elementary teachers will not see themselves as doers of science and will say that they do not have an idea of how science can be taught. Their noted goals often include: How to teach science in the classroom; how to present scientific information/learning in an exciting, engaging, effective way; and what methods, activities, and different experiments can make science fun!

The goal – to make science fun for their students – creates a serious dilemma, as their experiences with science have been, for the most part, anything but fun. In my elementary science methods course, the first assignment, a reflection describing how they think about science and their experiences with science, usually paints a picture of science as boring, arduous, textbook-oriented, based in the memorization of facts now long forgotten, impossible to understand, and meant only for smart people (of whom they are not, even though they already hold bachelors degrees with GPAs of a B+ or better). Many describe the college courses they have struggled through and failed. Most expect that the science methods course will laden them with additional fear, frustration, and failure. Few can remember any science in elementary school. Not many can connect science to their everyday activities even though they may raise animals or be avid gardeners or cooks. Some hold treasured memories of family visits to museums, walks in the woods, and the special teacher who made science fun. However, for the most part, what elementary preservice teachers have experienced as science lies on the opposite end of the continuum, far from the fun and engaging science they seek for their students.

On the first day of class I have the students draw a picture of a teacher teaching science to children. The instructor can follow this drawing with a metaphor/simile assignment with a starter of Teaching elementary science is/is like. .... This assignment helps them think more deeply about elementary science instruction. At the same time, the instructor needs to reflect upon what it is like to mentor new teachers as science teachers.

Together, as a class, the instructor can have the preservice teachers list their goals for their elementary science classrooms and what they want their elementary science classrooms to look like and sound like. Meanwhile the instructor should list his or her goals for the preservice teachers and the methods class. As a result of these activities, the preservice teachers and instructor begin to build an image of what their science teaching and the course should include.

At this point, the dilemma described above looms like an impending storm on the horizon – how does a preservice teacher develop science teaching practices that are both effective and fun and interesting for their students? How does a preservice teacher teach these lessons in school settings where science is often not taught? How does a methods instructor/professor assist beginning teachers in this quandary?

The Course as an Invitation: Becoming More Excited and Less Fearful of Teaching Science

With the beliefs and experiences of the preservice teacher in mind and my insights about the mentoring process, I present the elementary science methods class as an invitation to (1) construct a new view of science activity both as learners and as teachers and (2) develop personal identities as science doers and knowledge producers. The course is an invitation to explore science and science teaching, to learn something new, and for each individual to move along her/his personal/professional continuum with no delineated expectations of what the new/revised perceptions, knowledge, and skills will consist of except to achieve their stated goals. The mechanism I have used for extending the invitation is Guided and Open Inquiry using the 5 Es Learning Cycle (Atkin & Karplus, 1962; Bybee et al., 2006; Colburn, 2000). This is followed by productive reflection comprised of rich description in concert with critical analysis. Beginning teachers are asked to address specific issues and to focus on learners and subject matter, children and their ideas, and how learning takes place. Thus, they are asked to make linkages between learners and learning and the other aspects of teaching (Davis, 2006).

With these tools, we investigate: (1) their participation in science throughout their lives as learners and as teachers; (2) science concepts through everyday phenomena; (3) science teaching and learning in their elementary classrooms; and (4) their school communities. Early in the course, the inquiries are more guided; preservice teachers are given the question, problem, phenomena to investigate and the materials they need. We investigate Oobleck to learn more about the concept of properties, data collection, and the integration of language arts. We explore the moon and the seasons to learn more about the patterns of motion of celestial objects, methods to assess students' prior knowledge and misconceptions, and the social construction of new understandings. With batteries and bulbs, we explore circuits and energy and investigate student and teacher questioning.

As the semester progresses, the inquiries become more open; the topics for exploration belong to the preservice teachers. A bridge to open, student-driven inquiry is now needed. A class devoted to the investigation of the "pill bug" provides a transition to making observations, generating investigatable questions, designing investigations, collecting, recording, and analyzing data, and explaining findings. Following this initial, open-inquiry experience, preservice teachers, as members of cooperative groups, can engage in investigations of an overarching driving question, such as "Is it spring yet?" (Weiss, 1997).

Wrenches in the Mechanism: Issues of Access and Participation

To develop their practice, preservice teachers need to teach science to students (Anderson & Mitchener, 1994). So, the instructor can ask preservice teachers to teach at least two science lessons to children. Here, a serious roadblock to sharing science activity with elementary students usually appears. Little time, if any at all, is allotted for science teaching in most elementary school classrooms. Many schools focus on statewide achievement tests in reading and mathematics, and administrators have drastically reduced, or eliminated, science instruction at the preK-5 grade level. Also, most elementary preservice teachers struggle with how much instructional "say" they have and often must incorporate established policies and practices (Sadler, 2006).

To better understand the roadblocks, engage preservice teachers in a community exploration. In this assignment, preservice teachers prepare a narrative description of the community and school in which they are working that includes the following:

- A description of their classroom and school community including their students' interests, needs, and strengths, and the demographics of the teachers and students. Ask them to reflect upon how the teachers' demographics match the students' demographics and how similarities and differences impact learning and teaching, if at all. Ask them to interview one teacher and one student to find out about the school's strengths and its challenges.
- A description of the surrounding businesses, industry, housing, community services, libraries, museums, parks, nature centers, etc. to paint a picture of the school environment and teaching resources.
- A brief report of three different community interactions: (1) a visit to and observation of one of the community entities to learn more about what happens there that might add to their understanding of the community and inform their teaching, (2) an informal interview with one individual in the community to find out about the schools' strengths and the challenges, and (3) a community participation activity to learn more about the school and its neighborhood.
- A critical reflection about what they have gleaned from their study of the school and the surrounding community. Here they make linkages to their own background and school experiences. Also, they think about how the new knowledge they have acquired through the exploration influences their teaching and their relationships with the parents, teachers, and children in their school community.

Through guided inquiry and critical reflection of their teaching context, they begin to build an understanding of their school settings, the resources they do/do not have, and the challenges that they face. Such inquiry and reflection opens the door to talk about oppressive practices that squeeze individuals out of science or block their entry altogether. The preservice teachers in the methods class can then discuss how teaching is a political act and requires knowledge of: (1) the educational system within which one works, (2) the capital and power various individuals have, including themselves, (3) the challenges novice and experienced teachers face in the classroom due to inequitable social structures, and (4) the multitude of strategies that one can employ to address these challenges. This activity provides the instructor the opportunity to share experiences as an elementary teacher and as a university professor from a veteran's perspective and with a critical eye. Instructors can help preservice teachers interpret the settings in which they teach and how they might work through these systems and/or create change. The instructor also can connect preservice teachers with other educators who have knowledge of school settings, have capital within it, and can assist novice teachers in reaching their goals.

Leadership Response

What does it mean to me to lead/mentor new or experienced elementary teachers to science teaching? I value how this form of leadership keeps elementary science teaching real for me. As I work with preservice and experienced teachers, I reconnect with the joy and excitement that children express when engaged in science activity and with the delight of the flurry of questions they pose when asked what they want to know or what they think. I, like their teachers, are elated when students are so involved. The mentoring/leadership partnership also allows me to experience preservice teachers' expressions of wonder, uncertainty, and enthusiasm – recaptured from their youth – as they investigate their many questions through newly found inquiry.

Leadership through mentoring also enables me to remain close to the realities of schools, the daily struggles teachers face, and the ongoing myriad of required practices that are all deemed necessary in order to leave no child behind or to be highly qualified. I can better accompany the invitations to teach science with strategies to address the obstacles they experience. This keeps my relationships with preservice teachers honest, sincere, and productive.

Through these partnerships, I get to engage in my role as a change agent. As I reflect on the goal that brought me to science teacher education, I feel that some change has been enacted – enough to positively influence the science beliefs and experiences of many of the preservice and classroom elementary teachers with whom I have worked, and, through them, their students. Quite possibly, I have mentored some teachers to be change agents as well.

How have I used feedback from preservice and experienced teachers to become a better mentor and leader? I use multiple tools to provide me feedback about the ways the leadership activities described above influence preservice teachers as they develop science teaching beliefs and practice, and to give me a clear idea of what my preservice teachers are thinking about teaching and learning. Such feedback enables me to identify additional obstacles, concerns, and strategies so as to improve upon my practice; keeps my goal-directed behavior on course; and enhances my learning as a science teacher educator of what approaches are most helpful (London, 2002). I can get a sense of whether the preservice teachers have received enough in the way of support, challenges, and coaching to envision themselves as doers of science, knowledge producers, and science teachers. Early in my career I focused exclusively on the end-of-the-semester teaching evaluations, but this only provided information of what preservice teachers expected of the course and who they anticipated me to be as the instructor – important to consider, but not necessarily related to my goals.

Over the last decade, the feedback that I have received has led me to better make linkages between the inquiry that we did in the methods class and the preservice teachers teaching in their classrooms. I have learned to make more explicit their goals and my goals for learning and science teaching and to clarify how course activities were aligned with those goals.

The community exploration assignment was designed to enable preservice teachers to see the knowledge and skills of community residents, including children, and the goals and efforts for learning and success put forth in the neighborhoods where they teach as well as the obstacles to learning. I have come to see that it is difficult for many preservice teachers to see positive endeavors in cultural settings different from their own. It is also difficult for them to see the deep, underlying mechanisms within schools and society that foster impediments to good teaching and learning. This has led me to spend more time in school environments and to find members of the community to help me better understand these settings. By doing this, I can relate my learning experiences and encourage my preservice teachers to investigate their school milieu more carefully and listen to the stories and experiences of those who reside there. Spending time in their schools has enabled me to speak directly to their real situations and to assist them in planning for ways to teach science in effective ways.

In sum, my pathway through science teaching and learning has been rewarding. My leadership, expressed as mentoring new and experienced elementary teachers, has been fulfilling and has validated the personal excitement I feel with scientific inquiry and the pleasure that comes with children that are engaged in exploration and invested in their learning. I now have many colleagues who are willing to teach science and with whom I can collaborate to create change in schools so that more children can have access to science learning.

References

American Association of University Women (AAUW). (1989). *The Eleanor Roosevelt Fund for Women and Girls.* [Brochure]. Washington, DC: AAUW Educational Foundation.

Anderson, R. D., & Mitchener, C. P. (1994). Research on science teacher education. In D. L. Gabel (Ed.), *Handbook of research on science teaching and learning: A project of the National Science Teachers Association* (pp. 3–44). New York: Macmillan Publishing Co.

Atkin, J. M., & Karplus, R. (1962). Discovery of invention? *Science Teacher,* 29(5), 45.

Borko, H., & Putnam, R. T. (1996). Learning to teach. In R. Calfee and D. Berliner (Eds.), *Handbook on educational psychology* (pp. 673–708). New York: Macmillan Publishing.

Bourdieu, P. (1986). The forms of capital. In J. G. Richardson (Ed.) *Handbook of theory and research for the sociology of education* (pp. 241–258). New York: Greenwood Press.

Brown, L. M. (1991). A problem of vision: The development of voice and relational knowledge in girls ages seven to sixteen. *Women's Studies Quarterly, 19*(1 and 2), 52–71.

Browne, S. E. (no date). *Philosophy of mentoring.* Unpublished manuscript.

Bybee, R. W., Taylor, J. A., Gardner, A., Van Scotter, P., Powell, J. C., Westbrook, A., et al. (2006). *The BSCS 5Es instructional model: Origins, effectiveness, and applications. Executive summary.* Colorado Springs, CO: BSCS.

Colburn, A. (2000, March). An inquiry primer. *Science Scope, 23*(6), 42–44.

Cullen, D. L., & Luna, G. (1993). Women mentoring in academe: Addressing the gender gap in higher education. *Gender and Education, 5*(2), 125–137.

Davis, E. A. (2006). Characterizing productive reflection among preservice elementary teachers: Seeing what matters. *Teaching and Teacher Education, 22*(3), 281–301.

Davis, E. A., Petish, D., & Smithey, J. (2006). Challenges new science teachers face. *Review of Educational Research, 76*(4), 607–651.

Davis, K. S. (2002). Advocating for equitable science learning opportunities for girls in an urban city youth club and the roadblocks women science educators face. *Journal of Research in Science Teaching, 39*(2), 151–63.

Davis, K. S. (2001). "Peripheral and subversive": Women making connections, challenging the boundaries, and defining the limits of the science community. *Science Education, 85*(4), 368–409.

Davis, K. S. (1999). Why science? Women scientists and their pathways along the road less traveled. *Journal of Women and Minorities in Science and Engineering, 5*(2), 129–153.

Davis, K. S., & Falba, C. J. (2003). Integrating technology in elementary preservice teacher education: Orchestrating scientific inquiry in meaningful ways. *Journal of Science Teacher Education, 13*(4), 303–329.

Delamont, S. (1989). *Knowledgeable women: Structuralism and the reproduction of elites*. New York: Routledge.

Harding, S. (1991). *Whose science? Whose knowledge? Thinking from women's lives*. Ithaca, NY: Cornell University Press.

Hurd, P. (2000). *Transforming middle school science education*. New York: Teachers College Press.

Lave, J., & Wenger, E. (1991). *Situated learning: Legitimate peripheral participation*. Cambridge, MA: Cambridge University Press.

London, M. (2002). *Leadership development: Paths to self-insight and professional growth*. Mahwah, NJ: Lawrence Erlbaum Associates.

Luke, C., & Gore, J. (1992). Women in the academy: Strategy, struggle, and survival. In C. Luke and J. Gore (Eds.) *Feminisms and critical pedagogy* (pp. 192–210). New York: Routledge.

National Research Council. (1996). *The national science education standards*. Washington, DC: National Academy Press.

Packard, B. W. (2004–2005). Mentoring and retention in college science: Reflections on the sophomore year. *Journal of College Student Retention Research Theory and Practice, 6*(3), 289–300.

Packard, B. W., & Nguyen, D. (2003). Science career-related possible selves of adolescent girls: a longitudinal study. *Journal of Career Development, 29*(4), 251–263.

Packard, B. W., Walsh, L., & Seidenberg, S. (2004). Will that be one mentor or two? A cross-sectional study of women's mentoring in college. *Mentoring & Tutoring, 12*(1), 71–85.

Roychoudhury, A., Tippins, D. J., & Nichols, S. E. (1995). Gender-inclusive science teaching: A feminist-constructivist approach. *Journal of Research in Science Teaching, 32*(9), 897–924

Sadler, T. D. (2006). "I won't last three weeks": Preservice science teachers reflect on their student-teaching experiences. *Journal of Science Teacher Education, 17*(3), 217–241.

Stanton-Salazar, R. D., Vasquez, O. A., & Mehan, H. (1995). Engineering success through institutional support. In A. Murtado (Ed.) *The educational achievement of Latinos* (Vol. II). Santa Cruz, CA: Regents of the University of California.

Varanka-Martin, M. (1997). *The fate of an all girls' physics class: A reflection of society*. Paper presented at the Annual Meeting of the National Association for Research in Science Teaching, Chicago, IL.

Weiss, T. H. (1997). Has spring sprung? *Science Scope, 20*(7), 8–12.

Young, I. M. (1990). Polity and group difference: A critique of the ideal of universal citizenship. In C. Sunstein (Ed.) *Feminism and political theory* (pp. 245–247). Chicago: University of Chicago.

Chapter 11
Laying the Ladder Down: Egalitarian Leadership

Meta Van Sickle and Diane Cudahy

This chapter presents the story of Meta, an egalitarian leader who is growing to understand her leadership style and the life forces that contribute to its development. These forces include, but are not limited to faith and theology (Hoare, 2000), the ethic of care (Noddings, 1992), transformative learning (Mezirow, 1997), and nature of science (NOS) (Lederman & Lederman, 2004; McComas, 1998). This ethnographic study of an individual revealed that ways of knowing and operating that once were intuitive have become deliberate. These forces are belief structures which guide Meta's decisions as she makes career choices and are afforded/accepted multiple opportunities to lead.

An anthropological study of self, designed to describe and identify this leadership style, was uncomfortable from its inception. Meta's science self said it was biased to tell her story alone. Her Quaker self would comment that she "hadn't consulted the community." Both observations were correct and unacceptable to her. Thus, Meta invited Diane to participate as an interviewer and co-author of the chapter. Diane's responsibility was to ask probing questions, authenticate answers, mirror responses, and seek clarifications, in a collaborative journey of understanding of Meta's development. As a coauthor, Diane used the third person's voice except for the places where she quoted Meta's exact words. As a reader you will know this when you encounter quotation marks.

In this chapter we explore Meta's development as an egalitarian leader, identify theoretical frameworks that support her egalitarian practices, describe the positive and negative aspects of leading from this perspective and offer a thinking map for those who wish to further explore this type of leadership.

Egalitarian Leadership

Little is written about egalitarian leadership. Thompson (1994) posited that egalitarianism shares two components: sameness and equality. He described sameness as a belief that people are racially and culturally the same. He believed that because

M. Van Sickle (✉)
Department of Teacher Education, College of Charleston, Charleston, SC, USA
e-mail: vansicklem@cofc.edu

K.C. Wieseman, M.H. Weinburgh (eds.), *Women's Experiences in Leadership in K-16 Science Education Communities, Becoming and Being*, DOI 10.1007/978-90-481-2239-4_11,

both race and culture are constructions of humans then we differ in our definitions, not the concepts. Sameness resides in the human response and all are able to respond. Egalitarian leadership builds on this sameness. Equality represents equal access for all in political, economic, and legal life. Egalitarian leadership therefore is based on the notion of equal distribution of power in the community or group. It is not held just by the leader, or by a few in political power, neither is it sought.

Ashkanasy (2007) described egalitarian leadership as a style that was generous and compassionate while being group-oriented and focused on building a collaborative team. He also suggests that such a leader needs to be honest, sincere, and modest. There is an emphasis on team, integrity, and modesty. Egalitarian leadership stands in stark contrast to traditional authoritarian leadership style which Lipman-Blumen (1997) described as competitive, and ruggedly individualistic leadership and was based upon the assumption that the leader knows everything and knows what is best for the organization (Brennen, 1996). It also differs in the fact that an egalitarian leader does not solicit data to support his or her already made decision, but instead will always stand down from his or her own decision or opinion and follow a group decision. If the organizational structure demands such, an egalitarian leader will even bring forth the decision of the lower group to the higher ups.

Discerning Why and How She Leads

As we sought to understand and describe Meta's leadership style and the forces that shaped it, we embarked on a collaborative journey. A picture of a female science educator who leads as she lives, as an integrated egalitarian person, emerged from data gathered during countless hours of observation, probing questions, authenticating answers, mirroring responses, and seeking clarifications. Diane checked with at least two-thirds of current academic department members and many others who have worked with her on committees to ensure accuracy in reporting.

As an egalitarian leader inside and outside of the classroom, Meta uses the same practices in all walks of life whether she is dealing with a board of directors or a fifth grade science class, the church choir or her college classroom and even the writing process to generate this chapter. She has come to realize over the years that she is much more efficient, much more effective, and much more whole when operating in all arenas from a set of egalitarian principles. She does not call herself a leader even as she leads. She does not see herself as a leader even as others do. She speaks of facilitating, organizing, and directing as a choir director directs the choir, but the word leader is not one that she uses often. Meta knows that she will only accept a leadership position if she is free to operate from this style. When she tries to operate in traditional forms of leadership the experience is so uncomfortable that it literally takes a physical toll on her.

In Relationship to Her Faith

Being raised in the Quaker faith has a profound impact not just on what this leader spiritually believes, but on the way she envisions her role, works with people, and "runs" a meeting. The Quaker meetings she attended as a child with her grandmother were true to the Quaker tradition in that the meeting was held mostly in silence, people spoke only when "led" to do so, and men and women were treated equally in terms of "calling." Titles and "leaders" were not used, but the encouragement of "gifts" was. It was as likely that a man or a woman would be the "clerk" of the session. It was as likely that the spirit would guide a woman to speak as often or as easily as a man. Social issues were dealt with after the meeting at coffee hour until a group consensus was gained.

Her community-centered faith has at its core egalitarianism. All are equal and have potential. All bring "gifts" to the table. This is how she views people in groups she leads. Everyone is able to bring information forward and having done so, a decision is made and then carried out even if an individual does not agree. Leadership in her culture and Church is by consensus and the collective. So too is how she leads in academia. She has internalized the Church's belief that leadership is not an individual act through which one forces their will or vision on the will of others and acts accordingly in all settings.

Meta was taught to "Be still and know." Literally at her grandmother's knee she learned to be silent, to be calm. Contemplative community within the Quaker tradition is used to allow space for people to listen first, learn and talk, and finally act. Bass (1997) describes the purpose and practice as

> Friends sit in silence together, listen to the movement of Spirit, speak out of the silence and thus minister to one another, seeking unity. The practice of simplicity walks hand in hand with shared practices of worship and discernment and is a manifestation of the desire to live in what Friends speak of as Truth revealed in fidelity to the guidance of the Inner Light of the Christ. Paradoxically, Friends have found that centering their life in Truth enable those to shed cumber, and that shedding cumber enables them to perceive divine order more adequately. (p. 56)

In silence and contemplation where the mind is able to free itself from clutter, calm down, and hear, comes understanding. In her rural Michigan farm community, children were not excluded from conversations, although they were expected to listen and be silent. Often she had to "learn the way" and that meant to sit back, listen, and think. Sometimes this meant sitting and "stewing" until she "got it." Then it was hers! Now, as a leader, Meta challenges those whom she leads to act in the same manner, to "sit and stew," to explore options, to articulate positions until they "find it." Meta works to create the silence in which she believes we must sit and is comfortable with this silence.

Meta learned the art of and basic differences between explicit and implicit questioning in community. As an adult she honed the skills of a reflective listener and learned to talk to discover questions and to acknowledge what she did not know in order to listen to those around her. Quakers would say ponder, in education-speak we say reflect.

There is deliberate use of this silence and questioning in her meetings. She guides through listening and questioning, not answering and telling. When she asks questions, they are open-ended, chosen after careful listening, to elicit further understanding. At times she asks members to think about a question far in advance of a meeting. She also publicly asks those with quick answers to give the others some more time to formulate their responses. She does not force a group into a decision and feels comfortable explaining to those who are demanding decisions that her group is not ready to make one at that time.

In the Quaker faith, it is also the notion that people in any group are Friends and not strangers. Strangers are taken in to become part of the group even though the tradition is not to proselytize. As a leader, Meta views each group with which she is associated as a community. Every member of the group belongs, is a Friend. Meta looks for the strengths of members and accepts peoples' skills for what they are. She knows that she does not have any need to "be in charge," so she is able to step back and even push others forward. She encourages others to lead when they are willing. When she does take the central role (and at times the business of an organization requires that she do this) and the task is accomplished, she folds back into the community. She absolutely means it when she says to her academic department, "I'll continue in the chair's job if the group wants me to." Because she does not operate from a position of personal power, she is quite willing and comfortable facilitating the will of the committee or department even when it is not her decision.

In Relationship to the Ethic of Care

The ethic of care shapes Meta's leadership development. Caring or an ethic of care as described by Noddings (1984, 1992) requires one-caring and one-cared for. This means that caring is relational and that the carer and the one-cared for each make important contributions to the relationship. The carer needs to attend to the cared-for and the cared-for needs to respond to the carer. The ethic of care assumes that the carer wants to do "right" or "good" for the cared-for. Teven and McCroskey's (1997) perspective clearly indicated that the more learners perceive their teacher cares about them, the more the learners will care about the class, and the more likely they will be to pay attention in class and consequently learn more course material.

Meta applies the ethic of care to her leadership by showing those she leads that she cares deeply about them. In turn they care more about the community, attend to discussion and each other and each other's ideas, even when disagreeing. In the classroom this translates to the teacher attending to the student to the degree and in a way that the student feels fully recognized and worthy of the interaction. Meta believes the same is true of adults working on committees or in departments to solve problems. They also want and need to be heard and attended.

As a leader, Meta often meets with group or community members one on one. She invites members to articulate their opinions, feelings, and reasoning and to test out ideas. The tenor of the community changes as she cares-for it. In response to being heard and attended, group members respond more freely, openly, and honestly. Group conversation becomes more polite, less hostile, and more attentive. Trust grows. Ideas are more freely shared. As one colleague stated, "I am not afraid to express myself, even when I think I am at that point in the minority opinion." Another said, "I always know she is listening and hearing!"

The space that is created in a cared-for community is space in which diverse ways of knowing, of expressing, of believing can enter and thrive. Diversity is welcomed by this leader, and eventually is viewed by the group as strength. Talking time is more readily and easily shared. People back off personal positions for the betterment of the group, as one colleague demonstrated when she said, "I am going to table my issue. That one is much more important at this time."

Meta often invites and/or persuades the cared-for to "trial" an approach. This happens individually or within the group setting. As the carer she works diligently to avoid coercion. If the cared-for feels coerced, then she allows him or her to express the hurt, listen again, and negotiate to a new solution (Noddings, 2001). The safe and responsive environment she creates by leading from an ethic of care fosters energetic dialogue with full participation.

While attending, reflection is needed. While reflecting there are at least four key ideas to address: diversity, imagination, not knowing, and joy (Cohen, 2001). Diversity means that ideas that are or are perceived to be different need to be aired. Imagination gives us the opportunity to explore and discover more about who we are as individuals or as a group. Acknowledging not knowing, and being able to sit comfortably while not knowing, opens us up to learning or discovery. Finally, joy motivates us, and touches us deeply. Learning more enhances our opportunity to experience joy. From reflection come creative solutions, views, and decisions. Meta encourages creative solutions that are not what is expected or even welcomed, and is willing to take these decisions up the chain of command. People under her leadership know that their ideas are heard and considered.

The ethic of care recognizes that in addition to reflection, the carer needs to be able to assist the cared-for identify the problem(s) to solve. As a leader, Meta helps others to look for signs of the problem, test out the ideas of what the problem might be, ask questions to help clarify the problem, seek as many answers to the problem as possible, think through the ramifications of the solutions, make choices about actions to take with the cared-for (not to "do to" the cared-for), and make changes. As the carer, she must be able to listen to the range of possibilities. Her capacity to tolerate dissonance and the ambiguity of multiple perspectives while giving all equal respect is evident in the ways she leads.

As a leader, Meta recognizes that if we fail to care-for the policies and practices that we use, we are chancing that they will become meaningless, trite, or openly oppressive. For example, in education many of us have been ashamed by practices that have caused or allowed colleagues to become sick, and policies that reduce teaching to "teaching to the test" or force the creation of classes for "happy" students who will fill up the seats. As a leader who operates from an ethic of care, Meta challenges those around her to care-for each other and to care-for their world of work. This is because caring-for refers to the direct, personal response of a carer for a cared-for (Noddings, 2001, p. 37). Caring about is important and can create the opportunities to care-for. Caring about allows us to "feel good" about our actions, but is empty unless we chose to care-for others.

In Relationship to Perspective Transformation

As a leader Meta asks and enables community members to examine their perspective privately and sometimes publicly so that they might see greater or different possibilities, make new connections, and/or embrace or suggest change. The process of effecting change in an adult's frame of reference has been identified as transformative learning (Cranton, 1994; Mezirow, 1997, 2000). This comes out of a theory of perspective transformation that Mezirow (cited in Imel, 1998) described as:

> The process of becoming critically aware of how and why our assumptions have come to constrain the way we perceive, understand, and feel about our world; changing these structures of habitual expectation to make possible a more inclusive, discriminating, and integrating perspective; and, finally, making choices or otherwise acting upon these new understandings. (para. 2)

A person's perspective or way of making meaning becomes more inclusive, tolerant, reflective, and accepting of others and others' ideas.

Meta leads others to examine habitual assumptions and ways of operating. In community, there is an opportunity to do this. When coupled with her science understandings of the nature of inquiry and how it takes time to think and develop a rationale for a position, explore ideas, or gather data, Meta is compelled to lead in a manner that generates cognitive dissonance in order to foster inquiry and receive the gifts and understandings of the community. Cognitive dissonance is when you see or hear an answer, explanation, or event that does not jive or fit with what you think you know. In times of cognitive dissonance, angst rises for some and joy for others. Angst and joy are both important to the process and can be experienced by a single individual. In response to a discrepancy some react in fear because their often unexplored set of beliefs and values feels as though it is under attack and has to be defended. Others experience the same discrepancy as an opportunity for growth and change. Cognitive dissonance is useful in transformative practice because truths are laid out and errors in fact or knowledge are challenged. Tough questions which require no immediate answers are asked. Often a group she is leading does not move to a vote (as required by the "system") until all have had the opportunity to examine their position, beliefs, and assumptions. She does not hurry others.

Meta believes that it is important to create cognitive dissonance as a way to change structures of habitual expectation and to help create a community of understanding. Meta uses these moments as times of mediation and negotiation, of learning about each other and caring for each other. In this space is time for added voices and unheard ideas, as others sit silently thinking about what they are feeling and others are saying. It provides an opportunity for members to transform individual perspectives, and often become more aligned with each other which results in a shared action. Also, as people examine their beliefs and assumptions, they become aware of why and how they disagree and are able to more easily articulate their thoughts without hurting the feelings of others.

A community member observed, "Meta models what we should all be doing, the way we should all be acting." An element of this leadership behavior is demonstrating a willingness to listen, learn, and change.

In Relationship to Science

At her core, this leader is a scientist. At the core of how she leads are a set of understandings that come from science. Meta has "the ability to use scientific knowledge to make informed decisions regarding personal and societal problems" that Lederman, (2006, p. 301) labels as science literacy. Making informed decisions necessitates knowing how these facts are determined within the sciences. This requires an epistemology or a study about the values and beliefs that are inherent when studying to gain or develop knowledge, an epistemology called nature of science or NOS. Many educators, including Meta, agree with these basic tenets in NOS: scientific knowledge is tentative and subjective; it involves humans as they infer, imagine, and create; and it has social and cultural influences (AAAS, 1993; Lederman, 2006). These tenets are easily identified in her leadership style. She models that decisions are durable but can be changed, should be viewed in context and with data, are based on meaning we construct, are made by people and have consequences.

The value and respect for scientific knowledge and process came to Meta at a very young age through her faith and rural farm culture. There was no separation between science, religion, and her way of life. Her church supports the notion that each human mind is finite and that in order to approach truth with a capital T, a group of minds must agree to work together. So it is in the global community of scientists as they engage in the continuous search for truth. Just as science is, in part, driven by inquiry and fueled by data from multiple sources, so is a community led by an egalitarian leader. She learned at an early age that success is partially determined by the interdependence of people who seek knowledge from multiple sources. On the farm these sources were oral history, farmers' knowledge, and scientists who presented the latest information. As a leader she gathers the group together and encourages members to "put their minds together" to inquire and seek solutions. Her goal is to create and protect an environment in which this can occur.

It is from her drive to make sense of our world (or a problem, or a decision, or a new policy) that Meta leads, and inquiry is at the heart of the search. There are at least four steps central to her actions when she leads: (1) What is the problem?

(2) What data need to be collected? (3) How/What evidence do I/we have to help answer the problem? (4) What is the best way to accomplish the task or answer the question based on the data? Data is critical in her ways of leading as it is in doing science. New data can lead to new interpretations and new truths. She helps and leads others to see that more than one viable explanation may co-reside at any given time. In a community, there can be multiple right answers. She is known for saying, "What is your evidence?" and "Show me the data."

To this end Meta shares every bit of information she has with a group she is leading. She seeks answers from multiple sources. She makes public all she finds or is told. She is totally transparent in the actions she takes. In her current position as department chair this means she shares with the members of the department the actions she will take. For example, agendas are open for all to direct, information from other meetings is shared with faculty in a timely manner, by the use of a list serve all department members are asked for information or ideas. Materials are given to participants before meetings so they can be read and thought about. After all of this has occurred, the group comes to an informed decision which directs Meta's next actions.

Leadership Response

There are benefits and liabilities of leading with this philosophical frame, both for those who are led and for the leader. We begin with the liabilities, many of which are the perceptions of others not within the community. This style is at times perceived as weak by those who fail to understand that it is a process that is deliberately chosen, planned, interceded, and worked. Because this style gives an appearance of conflict, it makes some individuals notably uncomfortable. Inquiry, consensus building, transformative learning, and the expression of individual truths take time. This is an unfamiliar model of leadership for many individuals. As one colleague said, "Stop giving me opportunities; just tell me what to do."

There is resistance to this type of leadership even in the community. People may speak louder in coercive tones. They may complain that an idea has already been discussed. They may ask for vote before the idea is fully considered in order to move on. If a meeting is missed, individuals want to start all over again, as if because they weren't present, the conversation didn't happen. Operating under egalitarian leadership is initially frustrating for some people when they first experience it, especially those who are used to quick decisions.

Taking time to find an answer acceptable to the group, which can be fully supported, often does not fit in with the "make a fast decision" work place and organizational structure. The complexity of our society and academic position require, in particular, more time and discussion, meanwhile we are often pressured for immediate results and decisions. Operating in the "still and calm" does not mesh easily with the lives and the demands on faculty members' time by others. This type of leadership rests heavily on the use of data, data that can be difficult to get, generate, or find. It is problematic to be transparent when those in the organization above are not operating in the same manner. The structure (short meeting times, too much to cover) can work against the process this leadership style is trying to ignite.

An egalitarian leader is highly challenged to carve out the space and time for silence and contemplation and build the relationships of trust on which success relies. The leader must be comfortable with ambiguity, messiness, unresolved issues, and receptiveness. Some people hold the notion that an egalitarian leader does not know what action to take or how to lead. Others accuse such a leader of being wishy-washy and trying not to hurt anyone's feelings. This egalitarian leader, Meta, has been accused of not being able to make a decision when in fact she is capable of and has easily made many appropriate crisis decisions.

Benefits abound from operating in an egalitarian leadership model. Egalitarian leadership celebrates and encourages diversity within the collective by making space for multiple opinions and many little truths. It allows for rancor in the ranks, more voices to be heard, processing on a sophisticated level, and forces constituents, whether they want to or not, to understand the big picture. There are multiple strengths in shared governance as a group seeks not the "right" answer but the best fit for the group at a given time. The decision making process builds trust and uncovers a willingness to try a new idea. This is because the group knows the decision can be revisited without hard feelings and changed because the group has new information, data, and insight.

Under this form of leadership all individuals have value. A person choosing this leadership style looks for strengths. The members of the community do not have to think alike and members are not limited to a common set of understandings. This diversity of opinions among individuals can lead to a synergy that yields extraordinary solutions. There is the likelihood of a consensus being generated when enough time is given. There is the likelihood of a more lasting type of solution to a problem to be developed when given time.

An egalitarian leader creates safety and calmness within a meeting. The leader conveys the notion that there are no strangers, that it is a community. This produces a reduction in animosity and competiveness within the group. No one in the group holds the best idea; instead the group seeks its best idea.

Leadership Activity: A Thinking Map for Other Leaders

There is very little written about egalitarian leadership in the literature. It is not a position Meta or others can read about, observe readily, or be "trained" to do. Rather it is a style that emerges from a set of beliefs, values, and experiences. From the data emerged a set of shared elements that can be found in this style of leadership. An egalitarian leader: (1) Empowers the group and its individuals instead of self; (2) Believes in and values every individual; (3) Requires and processes data from multiple sources and perspectives for decision making; (4) Understands the necessity and power of a contemplative community.

Table 11.1 Explore being an egalitarian leader

Trait	Response from this leader	Response from another
(1) Empowers the group and its individuals instead of self	To be an egalitarian leader means divesting oneself from one's ego and any arrogance telling one that one's ideas are the best. As a group processes an idea, the leader's opinion or position holds ONLY the same weight as any other individual. In order to lead individuals to thrive and grow, an egalitarian leader begins with a set of beliefs about the value of individuals and their abilities to discern problems and create solutions.	
(2) Believes in and values every individual	The community is essential to all the above because this is where multiple perspectives are generated, discussed, and analyzed, and new and more useful solutions based on data are formulated. An egalitarian leader cares-for the individuals as well as the community.	
(3) Requires and processes data from multiple sources and perspectives for decision making	The significance of the need for and use of data is evident all through our investigation. Data are sought, identified, shared, transparent, and reflected upon. The greater the diversity of data shared in a group, the higher the likelihood of a useful solution being generated. Without data there is no opportunity to gain new meaning.	
(4) Understands the necessity and power of a contemplative community	It takes its shape from a process that stems from the leader's (1) belief in the community and (2) that the community holds all the necessary strengths to "get the job done" and done well.	

In order to begin to become an egalitarian leader and fully utilize each of the four components suggested above, a person must first examine his/her belief structure in relationship to the four statements above to develop a thinking map to egalitarian leadership. Use the Table 11.1 to help you deeply reflect on each trait to see if you want to embark on the journey to become an egalitarian leader.

Once you have completed your responses above, next consider a metaphor for thinking about being an egalitarian leader. For example, we imagine egalitarian leadership as a horizontal model of leadership. In egalitarian leadership the traditional vertical model is laid down, creating a community of equals and placing everyone in the position of potential leader. When one "lays down" the vertical hierarchical ladder, all the parts remain, but are in a new position. Situating the ladder metaphor in a biological context the soil is under all, the rain can fall directly on all, the sun

Table 11.2 Your leadership metaphor

What is your metaphor? Use the space below to draw or write.

can get to each, and space is available for all. Then each grows as able and in time with space. The egalitarian leadership style is designed to move each and every person forward through their strengths (gifts). Many people lead when it is their time and their gifts or talents are appropriate. The insights from this chapter can assist a leader to lay the ladder down. What is your metaphor? (Table 11.2)

References

American Association for the Advancement of Science [AAAS]. (1993). *Benchmarks for science literacy.* New York: Oxford University Press.

Ashkanasy, N. M. (2007). The Australian enigma. In J.Chhokar, F. Brodbeck, and R.J. House (Eds.), *Culture and leadership across the world: The GLOBE book of in-depth studies of 25 societies* (pp. 299–334). Thousand Oaks, CA: Sage Publications.

Bass, D. (1997). *Practicing our faith.* San Francisco: Jossey-Bass, Inc.

Brennen, A, (1996). Article and resources on educational administration and supervision. *Leadership styles.* Retrieved February 29, 2008, from www.soencouragement.org/leadership-styles.htm

Cohen, J. (2001). Social and emotional education: Core concepts and practices. In J. Cohen (Ed.), *Caring classrooms: Intelligent schools* (pp. 3–29). New York: Teachers College Press.

Cranton, P. (1994). *Understanding and promoting transformative learning: A guide for educators of adults.* San Francisco, CA: Jossey-Bass.

Hoare, T. (2000). *Facts about Friends.* Retrieved on November 19, 2007, from http://www.quaker.org/friends.html.

Imel, S. (1998). *Transformative learning in adulthood.* Washington, D.C.: Office of Educational Research and Improvement. (ERIC Document Reproduction Service No. ED42326).

Lederman, N. (2006). Syntax of nature of science within inquiry and science instruction. In L. G. Flick & N.G. Lederman (Eds.), *Scientific inquiry and nature of science* (pp. 301–318). Netherlands: Springer.

Lederman, N., & Lederman, J. (2004). Revising instruction to teach nature of science. Modifying activities to enhance student understanding of science. *The Science Teacher, 71*(9), 36–39.

Lipman-Blumen, J. (1997). Connective leadership-a new paradigm. *Drucker Magazine, 1*(1), 12–19.

McComas, W. (1998). The principle elements of the nature of science: Dispelling the myths. In, W.F. McComas, (Ed.), *The nature of science in science education rationales and strategies* (pp. 53–72). Netherlands: Kluwer.

Mezirow, J. (1997). Transformative learning: Theory to practice. In P. Cranton (Ed.), New directions for adult and continuing education: No. 74. *Transformative learning in action: Insights from practice* (pp. 5–12). San Francisco, CA: Jossey-Bass.

Mezirow, J. (2000). *Learning as transformation: Critical perspective on a theory in progress.* San Francisco, CA: Jossey-Bass.

Noddings, N. (1984). *Caring, a feminine approach to ethics and moral education.* Berkeley: University of California Press.

Noddings, N. (1992). *The challenge to care in schools: An alternative approach to education.* New York: Teachers College Press.

Noddings, N. (2001). Care and Coercion in School Reform. *Journal of Educational Change, 2*(1), 35–43.

Teven, J., & McCroskey, J. (1997). The relationship of perceived teacher caring with student learning and teacher evaluation. *Communication Education, 46*(1), 1–9.

Thompson, E. (1994). *Fair enough: Egalitarians in Australia*. Sydney: Allen & Unwin.

Chapter 12
Building Leadership Capacity by Nurturing Community

Tamara Holmlund Nelson

Coming out of graduate school and dissertation writing, a lonely place, I looked forward to joining a professional community at my university and in my College of Education. Based upon my previous cross-disciplinary collaborations as a high school science teacher, I knew that professional collaboration with others outside my specific discipline could be rewarding and overcome the isolation of classroom teaching. I visited with the folks in the College of Science, making connections across our mostly varied but occasionally corresponding interests. As the only science educator in my department, I sought ways to collaborate across fields by building upon our common interests in teacher education and educational practices. However, I found working in higher education to be more isolated and isolating than high school teaching. People tended to live behind closed office doors and come together mostly for meetings that dealt with programmatic or procedural details rather than scholarly conversations. As such, I was disappointed to find the answer to my question, "Where is the community I can join?" was, with a few exceptions, "Neither in my university nor my college."

Experiencing an ever increasing feeling of loneliness in this profession of higher education, I sought other people with whom to collaborate. Rather than look within the university, I sought partnerships with science teachers and district, regional, and informal science educators. However, there are real and significant challenges to forming and nurturing partnerships across these differing contexts, including the differing foci and tempos of work (Brookhart & Loadman, 1990). Additionally, not only was I trying to identify who might be coparticipants with me in a community, my understanding of what I and others envisioned as community continued to grow. I moved from notions of partnership to a deeper understanding of what it means to coparticipate with others in a community of learners, where leadership is distributed across members. Finally, as I did find others who wanted to engage in similar work, the question of how people build and nurture community came to the forefront of my attention and effort.

T.H. Nelson (✉)
Department of Teaching and Learning, Washington State University Vancouver, Vancouver, WA, USA
e-mail: tnelson@vancouver.wsu.edu

K.C. Wieseman, M.H. Weinburgh (eds.), *Women's Experiences in Leadership in K-16 Science Education Communities, Becoming and Being*, DOI 10.1007/978-90-481-2239-4_12,

Conceptualizing Community

Partnerships between Schools and Universities

I began my search for community by establishing partnerships. Goodlad (1988) characterized partnerships as symbiotic relationships, with mutual interdependence and reciprocal benefits. Symbiosis can only occur between unlike entities. To foster interdependence, the members must have different areas of expertise to offer each other, and each must be willing to fulfill the needs of the other while attaining their own goals. In the case of schools and universities, partnerships form between groups or individuals coming from different cultures. Universities and schools have been different enough in "purpose, function, structure, clientele, reward systems, rules and regulations, ambience, ethos" to be "markedly different cultural entities" (Goodlad, 1988, p. 14). These inherent differences in the work of teachers, university faculty, and district or regional educators may contribute to conflict along multiple dimensions, including work tempo, reward, focus, and power (Brookhart & Loadman, 1990).

While institutes of higher education and local schools have been connected through teacher education since the mid-1800s (Herbst, 1989), these relationships have been fraught with discord from the beginning (Sirotnik & Goodlad, 1988). Historically, universities were quick to distance themselves from the preparation of teachers. Teacher colleges or colleges of education within universities assumed responsibility for teacher education, and a gulf widened between university faculty in the arts and sciences, teacher educators, and teachers (Herbst, 1989). This gulf led to the creation of hierarchical relationships, with power and prestige residing at the university level, tapering off in teacher colleges, trickling down in a small extent to high schools, and being nearly non-existent in elementary schools.

Even today, educational researchers tend to disdain teacher preparation (Darling-Hammond, 1994; Ginsberg & Rhodes, 2003; Teitel, 1994) and focus their work on describing problems with schools rather than "bringing knowledge to bear to act on problems" (Lieberman, 1988, p. 84). Teachers, in turn, view educational research as disconnected from the classroom, so remote that it has little applicability to their real needs. For teachers, legitimate research is that which directly connects to the real problems they face daily (Johnson, 1989). This disconnection between research and practice reflects the status differential between university faculty and teachers (Wiske, 1995). University knowledge traditionally has carried more weight than teacher knowledge. As a result, school–university relationships are characterized by university faculty coming into schools and telling teachers how they should teach. These relationships are usually university directed and often take a noblesse oblige form of service on the part of the university, assuming a "working on" rather than a "working with" disposition (Clark, 1988, p. 40). Sometimes teachers reject university assistance, slighting university knowledge and skills as valueless and irrelevant (Sarason, 1996; Snyder, 1994).

These historical and socio-cultural relationships raise the question: Can partnership activities between school and university educators foster the development of a community?

Defining Community

Grossman, Wineburg, and Woolworth (2001) asserted that the word community has been overused so as to become meaningless. They examined the complexity of defining community in an age of cyberspace, where people meet in virtual communities with little commitment to more traditional community norms such as shared values, trust, commitment, and "working together for the common good" (p. 2). Furman (2002) examined how the concept of an underpinning value of sameness that traditionally characterizes community might shift to one of difference in a post-modern world. I begin with a definition of community that builds on traditional norms such as shared values relevant to the reason for the community and a commitment to contribute to fulfilling the community's goals, yet I recognize that a community may include people who meet virtually and who bring very different life experiences.

In the past decade, recommendations for improvement in both classroom practices and in teacher professional development have emphasized the formation of communities of practice, communities of learners, professional learning communities, and communities of inquiry. Each has its own definition, yet all share a common focus on people coming together around a common purpose, sharing values about and commitment to this purpose, and building relationships through active participation and discourse amongst community members.

There are distinctions across these conceptualizations of community. The concept of a *community of practice* addresses learning amongst people involved in similar activities, in a work setting or school, for example. Brown and Duguid (2001) pointed out that inclusion in a community of practice may be due to participation in similar activities; for example, all employees manufacturing airplane wings or the teachers in a middle school teaching science. Lave and Wenger (1991) were more explicit in the associations amongst those in a community of practice, noting especially the relationships between experts and apprentices in a given situation. Within a community of practice, social interactions contribute to learning and identity formation as related to the shared activities. Most often, the term community of practice implies the development of shared values and norms and includes the potential for co-construction of understandings relative to the shared actions (Raelin, 1997). In critique of this, Jaworski (2006) pointed out that, especially with respect to improving educational practices, the development of shared values may perpetuate the status quo rather than challenge established beliefs and practices.

A defining characteristic of a *community of learners*, also referred to as a learning community, is that all members "participate in shared endeavors with others, with all playing active but often asymmetrical roles in sociocultural activity" (Rogoff, 1994, p. 209). A community of learners often addresses the fostering of collaborative relationships and discourse amongst students in a classroom (Brown & Campione, 1990). McLaughlin and Talbert (2006) studied and promoted teachers' learning communities as sites for learning about and improving classroom practices. Key to teachers' learning is collaborative reflection.

In the first decade of the twenty-first century, a significant amount of attention is being given to intentionally forming and supporting *professional learning communities (PLCs)* within schools and school districts (DuFour, Eaker, & DuFour, 2005; Hord, 1997; Little, 2003; McLaughlin & Talbert, 2006). Each component of the professional learning communities label underscores a significant aspect of the group: a focus on questions of a substantive professional nature; a concentration on coming to understand more about teaching and learning through collaborative action and reflection; and, a commitment to negotiating shared values, goals, and vision relative to a common purpose. Professional learning communities, in the DuFour and Eaker model (1998), demonstrate six characteristics: (1) a shared mission, vision, and values; (2) a collective inquiry with joint planning and coordinated action; (3) a collaborative team; (4) action orientation and experimentation; (5) continuous improvement; and (6) a results orientation. A primary goal for teachers' participation in PLCs is to disrupt the isolation of teaching and the privatization of instructional practices (DuFour et al., 2005).

A *community of inquiry* is characterized by reflection on "what is being or has been constructed and on the tools and practices involved in the process" (Wells, 1999, p. 124). Wells described how acts of metaknowing engage members in discussion, argumentation, and negotiation to construct collective understandings of situations at hand and in the future. When the community is composed of members who bring differing interests or roles, a reflective inquiry stance may allow "understandings that cross community barriers. The permeating power of inquiry overrides differences in roles and goals of practitioners in different communities" (Jaworski, 2006, p. 204). There is a distinction between a professional learning community adopting an inquiry approach in order to investigate some aspect of their collective practice and a group of people who adopt an inquiry stance toward the work at hand. Jaworksi (2006) described the former situation as the use of "inquiry as a tool" versus the latter stance as "inquiry as a way of being" (p. 187) or a collaborative inquiry stance. A learning community characterized by a collaborative inquiry stance is generative – members negotiate goals, ideas, and perspectives; share knowledge and learn from others' areas of expertise; co-construct a collective vision that can encompass the multiple contexts and perspectives of members; and critically reflect on their joint activity to continuously improve it and learn from it. From these activities new understandings and contributions to a knowledge base are created.

This brief review of four different types of professional communities cannot capture all the theoretical and empirical work being done around the notion of community. The community structures described above are those that have helped me consider possibilities for integrating rich professional growth with professional work. While there are distinctions across these various types of community structures and processes, it is quite likely that any group of people working together to achieve stated goals may exhibit, to a greater or lesser degree, some or all of these community characteristics. The labels professional learning community or community of inquiry can be applied a priori to a group, but it is only on closer examination of the relationships, actions, and discourse within the group that the nature of the community is revealed.

Leadership Activity: A Story of My Inquiry Community

I did find a group of people interested in forming a professional learning community, and we developed as a community of inquiry. In this section of the chapter I draw upon our development to illustrate practices and values that may help you develop a community of inquiry with others. In particular, I recommend the following actions:

- Nurture a communal stance of questioning and wondering.
- Co-generate an inquiry question to focus the community's attention.
- Be aware of the trade-offs between the time needed to nurture a community of inquiry and the time for project activities.
- Generate and give ongoing attention to collaborative norms.
- Be explicit about advocating for ideas/actions versus inquiring into possibilities for action.
- Distinguish discussion from dialogue and use each explicitly at appropriate times.
- Foster shared leadership.
- Practice individual and collective reflection on actions and decisions.

These aspects are italicized in the text as a cue to help you understand what the aspect could look like in action.

I now illustrate these key aspects and values for developing a community of inquiry by sharing examples gleaned from my participation in a professional community characterized by inquiry and learning. This community developed as a defined group of science and mathematics educators (dubbed the "Steering Committee" or SC) from a variety of institutions and worked together over three years to develop, implement, and understand a professional development project for secondary science and mathematics teachers in six school districts. Earlier I raised the question as to whether a partnership amongst people with differing work roles, experiences, pressures, and interests can develop as a community. In the SC, the twelve original members (as well as four who joined in later years) possessed a wide variety of expertise, beliefs, and knowledge in different areas. The SC was composed of elementary and middle school teachers, former teachers newly transitioning into science or mathematics professional development, district science and/or mathematics coordinators, university faculty, regional educational service district professional development providers, and state-level science/mathematics educators with well-established reputations in the field. Within a group holding such diverse professional roles, there was a high probability that issues associated with real or preconceived differences in status, experience, age, gender, or goal alignment could complicate the development of a professional community.

We began more as a partnership amongst people with differing areas of expertise than as a professional learning community or a community of inquiry. When people come together on a project, it is common to assign different tasks to different members and ask each to draw upon her/his particular area of expertise to accomplish some part of the project. In the first four months of working together we operated in this way. We had an immediate and pressing task: recruit lead teachers for the first year of the project, generate goals for the three-year professional development project, and plan a week-long summer institute for the lead teachers and designated administrators. As many groups do when they first come together, we generated a list of norms under which our group would operate. Some of these norms were procedural (co-generate agendas, take breaks, be on time, have equitable representation across stakeholders when making decisions) and some attended to interpersonal relationships (values associated with trust, respect, confidentiality, open communication). We initially used these norms as we planned and implemented the summer institute.

However, five months into the partnership we made an explicit decision to "launch as a professional learning community undertaking inquiry" (meeting notes, September, 2004). As the project involved helping teachers form PLCs, we felt it was important to experience the processes and structures ourselves – to develop an understanding of what challenges and possibilities might be encountered in developing a shared vision or working across time to better understand our practices as professional developers. A search of the literature makes it easy to identify characteristics that professional learning communities or communities of inquiry should exhibit. Less easy is actually nurturing those qualities amongst a working group of people. We *deliberately framed our joint efforts as collaborative inquiry* – i.e., Jaworksi's *inquiry as a tool* to understand. Over time, we sought to embody Gordon Wells' (1999) notion of *dialogic inquiry as a way of being*; we deliberately set out to "wonder, ask questions, and seek to understand" (p. 121) the professional development model we were creating and implementing.

Generating an inquiry or focus question is often recommended in the literature on collaborative inquiry and professional communities (DuFour & Eaker, 1998; Love, 2002). Once again, this is easier said than done. We began with the question: "What can a facilitator do to foster and sustain a culture of collaborative inquiry?" Asking this question led to negotiation about whether the true focus of our inquiry was ourselves or the teachers we were facilitating. In the next month, we modified our question to: "How can we foster and sustain a culture of collaborative inquiry?" Modifying the question reflected the varying interests across the group. Some members had a strong interest in understanding the nature of professional learning communities; others were more interested in the moves they could make as facilitators to help teachers develop effective learning communities. Members having different interests was understandable, as some SC members directly facilitated teachers' engagement in PLCs and others did not. While the difference between these conceptual versus procedural interests created tension, this was not necessarily negative. By employing a distributed leadership approach (Leithwood et al., 2007; Spillane, Halverson, & Diamond, 2001), the leader for a specific collaborative activity and the activity itself was shaped by the nature of the pressing need (e.g., a more process- or theory-oriented emphasis) in any given month. The renegotiation of our inquiry focus took considerable time yet also involved us in substantive dialogue that deepened our understandings of the project goals and our collective vision.

Over the year, we specifically *examined the trade-offs between devoting significant time to building a community of inquiry and needing time to do the extensive planning required to develop and implement* a professional development project for 22 schools. We raised questions about how we could function as both an inquiry community and as professional developers. Collectively we had a tremendous amount of experience as professional development providers and together we could determine responses to our procedural questions such as "How much information does an administrator need? Who should be responsible for this communication? How do we recruit more teachers for next year?" (meeting notes, December 17, 2004). On the other hand, we did not have immediate answers to questions about collaborative inquiry: "How can we sustain the conversation, the ideas, the connections, when we are apart from each other? What collaborative structures do we need to move forward on our inquiry? How do we deal with conflict? How do we ensure equity of voices?" (meeting notes, September 8 and October 19, 2004). These questions were challenging, and collaboratively addressing these questions was at times daunting (see Nelson, Slavit, Perkins, & Hathorn, 2008, for further discussion). Despite the tension between getting the project work done and inquiring into our project work, we did find significant value in the inquiry process and continued to pursue it throughout the duration of the three-year project.

One of the most valuable actions we took to enhance our transformation from a partnership to a community of inquiry was undertaking a book study on *The Adaptive School* (Garmston & Wellman, 1999). The choice of this book proved invaluable over time, particularly the chapter on *developing norms of collaboration*. These norms are significantly different from those we adopted earlier, and shifted the way we operated. Our attention became more focused on carefully listening to each other, paraphrasing and probing each other's ideas, noticing which ideas were on the table and which had been withdrawn, and presuming that we all came to the group with positive intentions even when emotional intensity was raised. We paid attention to *maintain[ing] a balance between advocacy and inquiry* (Garmston & Wellman, 1999) by using various conversational and text-based protocols that created space for questions, disagreements, and multiple perspectives to be heard. We assessed our group processes and climate with respect to these norms at least quarterly. We also collaboratively *selected one norm to focus on over a given time frame and reflected upon our ability to implement* it at the end of meetings.

Studying *The Adaptive School* also helped us *distinguish between discussion and dialogue*. Identifying these differences, which have to do with decision-making versus exploring ideas, helped us maintain the balance between advocating for an idea or action (which had great potential for raising the emotional temperature in the group) and setting aside advocacy to probe each other's beliefs, values, and knowledge about a given proposal. It took us time to internalize these distinctions, as evidenced by the following notes from a March 17, 2005, meeting:

> On group norms: we discussed the need to continue our discussion on norms, the questions that were raised at the last meeting and confusion over dialogue and decisions and the need to continue growing our understanding of how to effectively manage our meetings.

By our final retreat in June of 2007, we prefaced remarks with statements such as, "This is discussion not dialogue," or "I am taking that idea off the table so we don't have to have further dialogue about it." On many occasions merely mentioning the employment of a collaborative norm created space for humor and relieved building tension.

We also focused explicitly on *sharing leadership*. This focus was extremely important in our shift from a group of people partnering on a project to a community of inquiry. I do not mean to imply we were successful at sharing leadership, or that we transformed into a high-functioning community of inquiry in the first year or even the second. Nevertheless, responsibility for leading meetings, preparing activities, finding resources, and contributing to the group dialogue and progress was distributed across all members. Not forgetting that we also were doing professional development, all members of the Steering Committee took leadership roles in enacting activities with teachers.

Sharing leadership was not always easy and often surfaced the differences amongst members' roles and responsibilities. For example, three members of the Steering Committee (including me) were Principal Investigators and held ultimate responsibility for the enactment of the project activities. Another was designated as the Project Director. In the first few months of the project, prior to the identification of all who would join the Steering Committee, the four of us became known as the senior staff. This title not so subtly suggested a power differential within the group, and proved problematic over time. The eight other members of the SC facilitated teachers' PLCs, the senior staff did not. I will say more about this in the next section.

The efforts to establish shared leadership across our group emerged from our appreciation for each person's experiences and knowledge, and our recognition that without explicit attention some might develop a larger voice or presence in the group than others. The norms of collaboration, the use of protocols to structure idea sharing or reflection, and the *practice of reflection and dialogue* helped us develop trust and shared values across the group as well as a commitment to the co-constructed goals of the project activities. Even as group membership changed over the years of the project, these values were established as part of the group culture and not only survived, but strengthened over time.

Leadership Response: Responding to Challenges in the Community

I open my leadership response with a reflection on a distressing event, which shows the challenges in developing and sharing the leadership in a community of inquiry. The interpretations of the event are solely mine. While I have discussed my interpretations with everyone involved, I present the story as I experienced it. If you are questioning whether you can join with others to develop professionally and personally satisfying communities, know that we did not experience just one rocky situation. However, I give strong credit for the ability of our community to move forward from this event (and others) to our deep commitment to being a community of inquiry and our continuous employment of and reflection on collaborative norms.

As we moved into the second year of the project, each person faced new demands from their work outside the project. As people juggled new schedules and responsibilities and ultimately missed SC meetings, issues with time, communication, focus, and status differentials began to surface. Additionally, the project was becoming more complicated due to a three-fold expansion in the number of participating teachers. The distressing event illustrates how many issues collided and created a challenge that had the potential to destroy our community, but importantly, did not.

As context, it was previously mentioned that the four senior staff did not facilitate teachers' PLCs. Their responsibilities with respect to SC meetings included making room reservations, mapping out an agenda based upon all members' input, and assuming overall responsibility for making sure the group's decisions and agreed upon actions were carried out. Eight SC members (facilitators) directly worked with the teachers and facilitated PLC meetings in schools. Three SC members lived two hours away and drove once a month for Steering Committee meetings. The hardships of travel, the other commitments of people's jobs, and the increased time needed for facilitation all played a factor in numerous absences from meetings as the second year began. Interestingly, these absences not only raised concerns about commitment, they caused everyone to recognize the interdependence of group members, and the desire for all to be present at our meetings. We realized that our group processes had become so collaborative that we could not move forward with our inquiry without input from everyone. Also, due to the increased number of participating teachers, tension increased about the time devoted to dealing with details of grant implementation (accountability, evaluation), determining (and pursuing) a new collaborative inquiry focus, and providing time for facilitators to address issues and details particular to their work with teachers, aside from the collaborative inquiry.

To better accommodate the participation of all members, a decision was made to hold our December meeting via a telecommunications system. It was in the email correspondence for planning this meeting that tensions became explicit. Michael,[1] a facilitator who had volunteered to take the lead on part of this meeting, asked via email for feedback on two protocols to guide the SC through looking at data

[1] All names are pseudonyms

(observational field notes, PLC meeting agendas, and other text-based artifacts) from teachers' PLC meetings. (We often used existing or self-designed protocols as procedures to scaffold our conversations about data or text materials.) Pam, in conjunction with Valerie, (both senior staff), responded with questions and made modifications to the protocols. Pam stated:

> We would like an outcome from your piece to be that we will be able to create a poster summary stating where we think our lead teachers are at with their knowledge and beliefs about PLC work.... thanks, Michael, I do appreciate your contributions and the thoughtfulness with which you plan, Pam

In an email copied to all SC members Michael replied:

> I took the liberty of copying this to others. It seems appropriate, since the group is still negotiating some of these SC-inquiry related things... I'd really welcome comments from more folks, and I hope that I'll have time this week to react to the dialogue.

Michael's use of the word "dialogue" in the email is meaningful given our study of the difference between discussion (for advocacy and decision-making) and dialogue (for exploring ideas and understanding perspectives). We were also negotiating a new inquiry focus for year 2, and there was a consensus for the question, "How can we grow and support leaders in PLCs?" However, there were differing perspectives on what this question implied.

As Michael continued in the email, the first hint of a disconnection between his (facilitator) goals for the activity and Pam's and Valerie's (both senior staff) emerged:

> You'll see for the most part, I want to hold on to the protocols as first written. This focuses on our steering committee inquiry, but does not make explicit connections to program goals and timelines... I think that facilitators need to work on the *intrinsic* aspects of the work rather than the program management aspects... I trust that this work will powerfully inform all of us, whether we are facilitating or not.... It's interesting how something starts out seeming pretty straightforward – it's a lot of work just to design and facilitate 2 hours of SC collaboration. I think I've invested 6–8 hr so far, and there's more to do. On the upside I think I'm learning a lot. Respectfully, Michael

Michael was expressing some frustration with the time involved in planning an activity for the next meeting. I also perceived that he felt others did not trust his decisions. In my interpretation, his reference to "program management" expressed the growing frustration amongst the facilitators that too much time was spent on accountability to grant parameters to the detriment of addressing a pressing need to understand how to help a PLC develop. Finally, the statement "I trust that this work will powerfully inform all of us" perhaps reflected a perception that the senior staff held different interests than the facilitators.

Valerie responded and, after posing a number of questions about the protocols, ended with these comments:

> I was interested by your comment you think that facilitators need to work on the intrinsic aspects of the work (understanding PLC development and leader support) rather than program management.... I think we need to unpack this viewpoint – I do not really understand

> how you can separate the goals or learning targets from the development of people getting there. I'm not sure I get what you are saying – Let's continue the dialogue. Valerie

A comparison of Michael's and Valerie's words illuminates fundamental differences in their understandings of each other's perspectives. No one responded to these emails, an interesting response in my view.

The tone of the emails became worrisome to senior staff and, using email and face-to-face venues, the four senior staff discussed the seeming misunderstandings between one facilitator and two senior staff. At another location, three facilitators discussed the tension. Finally, John, a facilitator and also Michael and Ginny's supervisor, joined the conversation with the following email (excerpt):

> Valerie, I've just finished a meeting with Michael and Ginny where we reviewed your questions and concerns about the two protocols that Michael has proposed. Our discussion focused around why there seems to be a communication disconnect between our vision of the Looking at Teacher Work process and [the senior] staff's vision.... [facilitators'] primary concern at this point is to better assess where our PLCs are at so we can provide support they need to continue growing. As senior staff you are more interested in describing PLCs based on a predetermined set of criteria and collecting data based on those criteria. Because of these differing purposes I would like to propose that facilitators do the activity separate from senior staff and use Michael's original protocol. That protocol is designed to meet the needs that facilitators currently have.... Senior staff can do the activity as well if they feel it is valuable and use whatever protocol they feel will best meet their needs. Thanks for listening, John

These emails occurred over a four-day period, four days prior to the planned meeting. No other Steering Committee members joined the email conversation other than a senior staff member who began his email with the following: "John and others, let's not give up yet! I think there is still time for 'hearing' each other," and concluded with: "I would caution against a move that divides the membership of this community in such an important conversation."

Ultimately, the agenda for the December meeting proceeded as Michael had planned and the examination of data resulted in generating some ideas about teachers' knowledge and beliefs, as requested by Pam. For me, and I think I can say for most of us, it was another step in our inquiry process. We continued to raise more questions about the work we were doing, what we were learning, and what we still needed and wanted to understand. In relation to our community, we decided we preferred to meet face-to-face rather than via the telecommunication system.

As the email exchanges of the distressing event above demonstrate, sharing leadership was not always a simple process. Yet, in looking back I can see impressive successes and would describe our results as a culture of collegiality (Khourey-Bowers, Dinko, & Hart, 2005). One member who spoke infrequently in the beginning became a major contributor to planning and implementing the second summer academy. We became more equitable in selecting meeting sites. Our shared leadership trajectory led to discarding the senior staff role in overseeing the meeting agendas as well as an attempt to eliminate the senior staff designation. While the Principal Investigators still had the ultimate responsibility of fulfilling the grant requirements, the input of facilitators modified how information was collected. Many of the facilitators moved into leadership roles in other professional areas.

One of the most valuable decisions we made was to undertake the study of *The Adaptive School*. Our ongoing attention to and explicit reflection on the ideas in selected chapters, especially the norms of collaboration, helped us to incorporate many of them as cultural practices in our community. These norms provided a supportive framework for enculturating new members and addressing emergent emotional tensions. Rather than take comments personally, the admonition to "assume positive intentions" provided salve for what might otherwise sting.

While it was sometimes difficult to enact, we committed to probing each other's ideas and employing dialogue (rather than just discussion). We constantly reflected on this commitment, on what was working and what needed improvement in our interpersonal interactions. We paid attention to the consideration of others' ideas and tried to enact (and tolerate) a questioning stance. As shown in the distressing event, this wasn't always easy. Questioning sometimes was perceived as challenging or mistrusting another's ideas. Despite the ongoing attention to developing shared leadership, the ultimate decision-making power that resided in the Senior Staff (especially the PIs) sometimes pre-empted dialogue. While Senior Staff seldom intended for their ideas to be taken as the final word, inherent power differentials sometimes surfaced and closed down the dialogue. We addressed these differentials in various ways, including using protocols to ensure opportunities for all voices to be expressed, calling for a pause as we moved through an agenda so that perceived tensions could be explicitly examined, and rotating the decision-making process for meeting agendas among all members of the SC.

Consciously attending to ways for sharing leadership and the pursuit of our own inquiry were essential elements in our growth as a community of inquiry. We came together as a group of science and mathematics educators committed to working on a project we all believed would contribute to improving students' learning. Without an explicit focus on collaborative inquiry, I believe we would have remained a more traditional group of colleagues, each contributing expertise and learning new activities through our consultation. Because we each held our own visions of the support teachers needed, we could have adopted an approach in which we pieced together professional development activities based upon each SC member's ideas. Yet, in co-generating an inquiry question and continuing to devote time to this, we developed as a community of inquiry. Developing our inquiry stance increased our capacity as a group to question, share, and reshape our individually held ideas.

We truly wanted to learn from each other and learn together in areas where we held no expertise. Most of the SC had a real commitment to the inquiry stance of our community, and those less committed were willing to go along with it. Although we did not always understand many aspects of our collaborative inquiry – for example, how each was interpreting our inquiry question, what outcomes we wanted, or how best to proceed with our inquiry – it was through immersing ourselves in the process that we learned a tremendous amount about engaging in and supporting others in collaborative inquiry. Most importantly, we learned that distributing leadership helped to reveal the multiple ways in which members were viewing the work. As different people took on responsibility for moving the inquiry activities forward, differences in perspectives and values became apparent. It was critical to the growth of our community that we acknowledged the existence of these differences. Without doing so we would have been a loose coupling of professional development providers making individual contributions to the project. Committing to the sometimes uncomfortable and frustrating pursuit of understanding our professional development activities and impacts, we learned about fostering and sustaining a culture of collaborative inquiry for ourselves and for the teachers with whom we worked.

Generating an inquiry focus and pursuing it through shared or distributed leadership engaged us in substantive dialogue and contributed to our growth as a community of inquiry. We began by using inquiry as a tool and gradually moved toward inquiry as a way of being in our community (Jaworksi, 2006). Evidence of our desire to continue this community came during our last retreat in June of 2007 as we formulated plans to find the resources and time to come together in the upcoming year. Members stated that despite the ending of the grant we wanted to continue our dialogue about and inquiry into how to establish and support learning communities. The challenge was to find ways to continue our dialogic inquiry without the support of a funded project. Whether we are able to do this together, or move on to develop new communities of inquiry, I believe we all developed powerful approaches to leadership. For me, this experience is shaping the ways I engage with teachers and educators as we focus on the improvement of science teaching and student learning.

References

Brookhart, S. M., & Loadman, W. E. (1990). School-university collaboration: Different workplace cultures. *Contemporary Education, 61*(3), 125–128.

Brown, A. L., & Campione, J. C. (1990). Communities of learning and thinking, or a context by any other name. In D. Kuhn (Ed.), *Developmental perspectives on teaching and learning thinking skills* (Vol. 21, pp. 108–126). New York: Karger.

Brown, J. S., & Duguid, P. (2001). Knowledge and organization: A Social-Practice perspective. *Organization Science, 12*(2), 198–213.

Clark, R. W. (1988). School-university relationships: An interpretive review. In K. A. Sirotnik & J. I. Goodlad (Eds.), *School-university partnerships in action: Concepts, cases, and concerns* (pp. 32–63). New York: Teachers College Press.

Darling-Hammond, L. (1994). Developing professional development schools: Early lessons, challenge, and promise. In L. Darling-Hammond (Ed.), *Professional development schools: Schools for developing a profession* (pp. 1–27). New York: Teachers College Press.

DuFour, R., & Eaker, R. (1998). *Professional learning communities at work: Best practices for enhancing student achievement.* Bloomington, IN: National Educational Service.

DuFour, R., Eaker, R., & DuFour, R. (Eds.). (2005). *On common ground: The power of professional learning communities.* Bloomington, IN: The Solution Tree.

Furman, G. (2002). Postmodernism and community in schools: Unraveling the paradox. In G. Furman (Ed.), *School as community: From promise to practice* (pp. 51–75). Albany, NY: State University of New York Press.

Garmston, R. J., & Wellman, B. M. (1999). *The adaptive school: A sourcebook for developing collaborative groups*. Norwood, MA: Christopher-Gordon Publishers.

Ginsberg, R., & Rhodes, L. K. (2003). University faculty in partner schools. *Journal of Teacher Education, 54*(2), 150–162.

Goodlad, J. I. (1988). School–university partnerships for educational renewal: Rationale and concepts. In K. A. Sirotnik & J. I. Goodlad (Eds.), *School-university partnerships in action: Concepts, cases, and concerns* (pp. 3–32). New York: Teachers College Press.

Grossman, P. L., Wineburg, S., & Woolworth, S. (2001). Toward a theory of teacher community. *Teachers College Record, 103*(6), 942–1012.

Herbst, J. (1989). Teacher preparation in the nineteenth century: Institutions and purposes. In D. Warren (Ed.), *American teachers: History of a profession* (pp. 213–236). Washington, DC: AERA.

Hord, S. (1997). *Professional learning communities: Communities of continuous inquiry and improvement*. Austin, TX: Southwest Educational Development Laboratory.

Jaworski, B. (2006). Theory and practice in mathematics teaching development: Critical inquiry as a mode of learning in teaching. *Journal of Mathematics Teacher Education, 9*, 187–211.

Johnson, W. R. (1989). Teachers and teacher training in the twentieth century. In D. Warren (Ed.), *American teachers: History of a profession* (pp. 237–256). Washington, DC: AERA.

Khourey-Bowers, C., Dinko, R. L., & Hart, R. G. (2005). Influence of a shared leadership model in creating a school culture of inquiry and collegiality *Journal of Research in Science Teaching, 42*(1), 3–24.

Lave, J., & Wenger, E. (1991). *Situated learning: Legitimate peripheral participation*. Cambridge, U. K.: Cambridge University Press.

Leithwood, K., Mascall, B., Strauss, T., Sacks, R., Memon, N., & Yashkina, A. (2007). Distributing leadership to make schools smarter: Taking the ego out of the system. *Leadership and Policy in Schools, 6*(1), 37–67.

Lieberman, A. (1988). The metropolitan school study council: A living history. In K. A. Sirotnik & J. I. Goodlad (Eds.), *School–university partnerships in action: Concepts, cases, and concerns* (pp. 69–86). New York: Teachers College Press.

Little, J. W. (2003). Inside teacher community: Representations of classroom practice. *Teachers College Record, 105*(6), 913–945.

Love, N. (2002). *Using data/getting results: A practical guide for school improvement in mathematics and science*. Norwood, MA: Christopher-Gordon Publishers, Inc.

McLaughlin, M. W., & Talbert, J. E. (2006). *Building school-based teacher learning communities*. New York: Teachers College Press.

Nelson, T. H., Slavit, D., Perkins, M., & Hathorn, T. (2008). A culture of collaborative inquiry: Learning to develop and support professional learning communities. *Teachers College Record, 110*(6), 1269–1303.

Raelin, J. A. (1997). A Model of Work-Based Learning. *Organization Science, 8*(6), 563–578.

Rogoff, B. (1994). Developing understanding of the idea of communities of learners. *Mind, Culture, and Activity, 1*(4), 209–229.

Sarason, S. B. (1996). *Revisiting "The culture of the school and the problem of change"*. New York: Teachers College Press.

Sirotnik, K. A., & Goodlad, J. I. (1988). *School-university partnerships in action: Concepts, cases and concerns*. New York: Teachers College Press.

Snyder, J. (1994). Perils and potentials: A tale of two professional development schools. In L. Darling-Hammond (Ed.), *Professional development schools: Schools for developing a profession* (pp. 98–125). New York: Teachers College Press.

Spillane, J. P., Halverson, R., & Diamond, J. B. (2001). Investigating school leadership practice: A distributed perspective. *Educational Researcher, 30*(3), 23–28.

Teitel, L. (1994). Can school-university partnerships lead to the simultaneous renewal of schools and teacher education? *Journal of Teacher Education, 45*(4), 245–252.

Wells, G. (1999). *Dialogic inquiry: Towards a sociocultural practice and theory of education.* Cambridge, UK: Cambridge University Press.

Wiske, M. S. (1995). A cultural perspective on school-university collaboration. In D. N. Perkins (Ed.), *Software goes to school: Teaching for understanding with new technologies* (pp. 187–212). New York: Oxford University Press.

Chapter 13
I Don't Want to be a Principal! I Want to be a Teacher Leader!

Lisa Pike

Story and Theoretical Framework

After 15 years of blissful teaching, when encouraged the first time from my principal to pursue an administrator licensure program, my reaction was, You have got to be kidding me! Thanks for the vote of confidence, but no thanks. I really just want to share my expertise as an elementary teacher, and learn with and from other teachers. I had no interest in obtaining an administrator license, nor did I have extra time. In addition to teaching first grade full time, I was also extremely busy as a mother of two teenage athletic daughters, a wife, a 4-H leader, and an assistant coach of Middle School Cross Country! Apparently, my principal did not realize that when I said, Thanks, but no thanks, I actually meant it! I guess I was somewhat to blame for his misinterpretation. Over the years, I had built my own reputation of having difficulty saying no to just about anything, a character trait common to teacher leaders. Throughout that year, brochures, pamphlets, and fliers of all types mysteriously began to appear in my mailbox of different distance learning programs I could pursue to obtain the infamous administrator license! I must admit, I was a bit honored that he had such confidence in me to be an administrator, but that type of leadership was not what I had in mind. Enough was enough, and no finally meant no, but my administrator knew that would not be the end of my leadership journey! In May of that year, he left another gift in my mailbox, a pair of maracas, so that in the very near future, I could use them as a tool to facilitate meetings!

My desire for affecting change in our school became more powerful while working on my master's degree from 2000 to 2002. One assignment, among the many that stand out, was an educational analysis of Linda Darling-Hammond's *The Right to Learn*. In it, Darling-Hammond (1997) proposed a progressive blueprint for creating schools that work. What a concept! As I reflected on her suggestions, it hit me, quite literally, in black and white as I typed the words, I must step up to the plate and offer support and assistance to others. If I remain silent, knowing that just my students

L. Pike (✉)
Gunnison Community School, Gunnison, CO, USA
e-mail: lpike@gunnisonschools.net

K.C. Wieseman, M.H. Weinburgh (eds.), *Women's Experiences in Leadership in K-16 Science Education Communities, Becoming and Being*, DOI 10.1007/978-90-481-2239-4_13,

are getting it, then I am just as guilty as those who are ignorant and afraid of change. Those words proved to be the baby steps in my walk as teacher leader today. It also struck me that my newfound realization was simply not enough. Action was needed to ensure great strides in education. So I became more involved in curriculum work, professional development opportunities, research, and basically anything to extend my pedagogical knowledge within and beyond the classroom, in order to share it with others. Educational leadership skills remain the same throughout the disciplines, but the nature of the leader's knowledge is different. Leaders in the primary grades, such as myself, need less science content knowledge compared to leaders strictly in science education, because we must know a little about all content areas since we teach all of them.

Another step along my leadership path during this time included presenting research at local, national, and international levels. I was only a first grade teacher, what could I possibly offer to education at a higher level? While others believed in me, I still had my insecurities. When the presentations were well received, my self confidence quadrupled. For a few years, this was fulfilling. After all, I was already so busy with other things.

In 2004, I experienced professional restlessness (Danielson, 2007). The teacher education program at our local college received a large five-year grant that included building a partnership between the college and the school district. Paid through the grant, I became the Mentor Coordinator for our district, which involved redesigning our mentor program. This would be in addition to teaching full time. My work would be valued, after all, there was a price tag attached in the form of a stipend! This price tag also led to my first uncomfortable feelings of I am getting the hairy eyeball from other teachers in the district.

Historically, there has been somewhat of the Chosen Ones mentality from some resistant colleagues and even so-called friends. Some felt that the same people were always chosen for golden opportunities. Since I am a firm believer in the mindset of positive things happen to positive people, could it be they were chosen because they were the positive ones? As Mentor Coordinator I worked with a small team of teachers, administrators, and college faculty to redesign our school district's mentor program. This became what I like to call my first true leadership experience, in that our work resulted in a district-wide change, rather than just change within my own classroom. As part of this leadership I officially became the instructor of record for my first graduate level teaching experience. Teaching a course on mentoring as part of this work was an opportunity for some of the teachers to question my qualifications, experience, and expertise, a situation I perceived to be an issue of power. After all, I would now be in a position to assign grades to colleagues who were enrolled in the course. At first, the old school belief of the teacher/learner relationship with teacher being at the higher level was difficult for some of my colleagues to shake.

Until the redesign of the mentor program, the mentors in the district had never received any type of professional development in mentoring. Having been a mentor for new teachers numerous times in the past, I knew I had not been provided any type of mentor training essential to being an effective mentor. As the coordinator of this program, I decided to provide mentors the training they would need to be more effective in this role. Initially, a trainer was brought in from outside the district, which comforted the teachers who perceived outsiders to be the experts. I also was able to build and refine my mentor skills as a participant in repeated training sessions. After two years however, it became my role to provide these trainings. I felt very qualified and comfortable in providing mentor training. Just because I felt comfortable did not mean the mentors felt comfortable with me. Some were older or had more years in the district and privately asked, "How could Lisa possibly be the expert in this?" Relationships appeared normal on the surface; however, I sensed feelings of resentment on the part of some of the veteran teachers. I never questioned my abilities, but was saddened that others did.

Although the mentor program is now allowing for more effective mentoring in the district, some teachers perceive it as just more paperwork and I am to blame because mentors are now held accountable by documentation in the form of the mentor handbook. The new teachers, however, appreciate the support they receive and value the program as it is today.

While working on the redesign of the mentor program, I also became more involved in our district's efforts in curriculum alignment to the standards. I loved this process, knowing that collectively, our teaching would be more direct and explicit, and ultimately more effective. After I worked on unpacking the state science standards (taking the broad state standards and making them more useable), many colleagues thought I was crazy, volunteering to unpack language arts, math, and social studies standards as well. My grade level team-mates, however, were thrilled! They had no desire to take this on, and it was work that I was truly passionate about! After days of having substitute teachers in my classroom during the 2005–2006 year (pertaining to Mentor Coordinator position or curriculum work) I wondered if my colleagues perceived the travel and time away as fringe benefits, and if so, did they resent these perceived benefits? The fact that they did not resent the work was monumental!

When our district began curriculum work in 2005 with Jane E. Pollock, the director of the education consulting firm Learning Horizon, Inc., we came to realize the need for a tighter curriculum and clearly defined learning targets. Pollock has expanded upon what had been previously referred to in education as Curriculum, Instruction, Assessment, and Feedback to an approach she labeled as the Big Four to improve student learning (Pollock, 2007). The first tenet of the Big Four reminds us that it is not enough to simply have a curriculum. The curriculum needs to be well articulated, meaning that it includes clearly defined learning targets, and it needs to be used. Next, teachers need to plan for delivery, rather than merely instruction. Planning for delivery includes utilizing instructional strategies that are effective. In addition, teachers need to not only assess, but differentiate the assessments for all learners. Finally, feedback alone will not improve student achievement. The feedback needs to be specific to the learner based on the learning targets.

As Janie Pollock worked with small groups of teachers each time she visited our district, she helped us understand the importance of the Big Four and quite literally in *Improving Student Learning One Teacher at a Time* (Pollock, 2007). As I came to be one of these teachers, I also was able to help facilitate this process with other teachers in the district. We looked closely at our own classroom instruction, and quite frankly, for many of us it was a matter of classroom instruction that did not work! It was not until a small group of us began to meet, discuss, and try out the research based strategies in *Classroom Instruction that Works* (Marzano, Pickering, & Pollock, 2001) that we saw results! At this same time, our dynamic new Director of Curriculum and Assessment came on board and her vision of shared leadership provided professional development at each grade level. Teams were created to clearly define our learning targets by unpacking the broad state standards. This fostered buy in of teacher created documents. Her leadership was instrumental in making our district standards real. Today, the content standards notebooks in our district no longer sit upon the shelves collecting dust as they had previously, but are becoming worn out, beat up documents used to drive instruction in many teachers' classrooms. Each year, teachers in the district join the ranks of the CITW (Classroom Instruction that Works) group, utilize the strategies that work, and begin to see the need for assessments that clearly align to the standards and benchmarks. As Marzano et al. stated, "individual teachers can have a profound influence on student learning even in schools that are relatively ineffective" (2001, p. 3). That is precisely how I felt! By helping to create a clearly defined set of learning targets to impact student learning, I scratched my leadership itch. . .for awhile.

In the spring of 2006, I attended an overview on Response to Intervention (RtI), a three-tiered model of instruction for all students, matched to their needs. I was, of course, interested to learn more about this model with a team of teachers from my school. No surprise, it is the life-long learner in me. It just made sense that this would be the next step, after all of our work with standards, to ensure a strong core curriculum for all students. Shortly thereafter, the grant team from the college decided to fund an RtI Coordinator to facilitate the development and implementation of this model in our district. The job was posted as a Teacher on Special Assignment position and needless to say, I jumped at the chance. This was everything I believed in as an educator! I do admit I was a bit nervous. After all, at that point in the spring of 2006, I had been teaching in the classroom for 15 years, but this could finally be my chance to take the next step in my teacher leadership journey! I had to give it a whirl, despite even a lack of support on the home front. My husband could not understand why I would want to leave my first grade classroom and assume a coordinator position. The lack of support extended to include some of my colleagues as well. Even though I had to apply and interview for the position, some colleagues still maintained the Chosen One attitude, even going so far as to verbalize that this was strategic and the powers that be know who they want. I did not even have the job and I could not believe the rumor mill!

I was so excited when I was offered the position, yet I was disheartened that a few of my own colleagues and even friends would have this opinion! Was it jealousy? Did they secretly want the position? I was always the people pleaser who got along well with everyone, which, I am sure, was a direct carryover from others' perceptions of me as a person who plays well with others in elementary school. How would I face this dichotomy? I avoid conflict and tension at all costs and here I was contemplating this position. Was it even worth it? On one hand I knew, professionally, it was the right thing to do. On the other hand, did I want to be the recipient of negative vibes? I remembered what I wrote during my master's program about remaining silent. By accepting the position, these questions were answered, but my feelings were not resolved. However, different questions were soon to follow. Did anyone else in this type of role feel this way? Little did I know I would explore this question soon enough by writing this chapter.

In assuming my new roles as RtI/Mentor Coordinator at the district administration building, suddenly I was up against more uncomfortable feelings. I got the impression that folks perceived my new role as crossing over to the dark side. "Oh, you are over at the Head Shed now, you are one of them," many taunted. Even from one administrator, I heard, "Face it. You are an administrator now, Lisa." You have got to be kidding me! I had no power, no control, or no evaluative authority whatsoever! Not to mention, my pay check did not reflect an administrator's salary! It was true I still received a stipend, but with that stipend came many extra weeks of work in the summer, many after school and evening meetings, as well as teaching a graduate level class.

I will never forget the first days back to school in my new leadership role as RtI/Mentor Coordinator in August 2006. During a district-wide professional development inservice held at my former elementary school, I went to collect my mail from my mailbox only to discover that my name was gone and replaced with another! I felt alienated. Until that day, I had been able to overcome the negativity of the naysayers, yet I fought back tears when I was told I did not really need a mailbox, teachers could just send things over in interoffice mail. To me, the mailbox represented my continued membership in the teachers' community, not just being a leader and member of the Head Shed. I had to change my attitude and the only way to do this was to accept others' perceptions and realize that these were my issues, over which I alone had control.

The RtI Coordinator position was new to the district. Even though I had great mentors in the people serving as the Curriculum and Assessment Director and Director of Special Services, everything was new. No one had previously held this position and there were not many human resources nearby. I was not entirely sure of what I was doing and that felt odd, but I was up for the challenge and knew I would be successful. Ongoing, job-embedded professional development was a must and, thankfully, greatly supported.

"Is this really all in my job description?" I asked myself and others. I soon realized that Response to Intervention is a very, very large umbrella! The Colorado Department of Education in July, 2007, defined Response to Intervention as:

> an approach that promotes a well-integrated system connecting general, compensatory, gifted, and special education in providing high quality, standards-based instruction/ intervention that is matched to students' academic, social-emotional, and behavioral needs. A continuum of intervention tiers with increasing levels of intensity and duration is central to RtI. Collaborative educational decisions are based on data derived from frequent monitoring of student performance and rate of learning (e.g., Colorado Department of Education, 2007, Definition section, para. 1).

Without a shared leadership approach, this would be a daunting role.

Thankfully, the vision of shared leadership became highly regarded in our district at this same time. With all of our work with the Big Four and our focus truly on all students by implementing an RtI model, we were finally moving in the right direction. In the summer of 2006, our district began the implementation of the professional learning community (PLC) model set forth by DuFour, DuFour, Eaker, and Many (2006). As RtI Coordinator, one of my roles was to help plan the professional development that would be needed to facilitate the implementation of Response to Intervention. Working with the Directors of Special Education and Curriculum and Assessment, we merged the professional development into a four-day optional symposium before the start of the school year for over sixty teachers in the district. During 2006–2007 leadership teams comprised of teachers and an administrator at certain sites were created and began to function as PLCs. Slowly, these teams began to use common language in a collaborative culture! They learned how to establish and follow norms for meetings! They began to use student assessment data to create SMART goals for instructional decisions! Eventually these PLCs became more effective than teacher groups of the past! Agendas, norms, and meeting protocols were set and followed, and results were made with students and colleagues! A shared leadership approach allowed for the spread of responsibility throughout a leadership team. I did not have to be the expert! Collectively, we, as teams, became the experts!

In building a collaborative culture of leadership teams, Marzano, Waters, and McNulty (2005) offered two generalizations. The first generalization was that members should be volunteers, sharing a culture of commitment. We had that; we were not lacking for teachers who were committed to the process and who were also willing to step up to the plate. The second generalization was that members need to establish "strong operating principles and agreements" (p. 104), and in this way their work together is productive. By setting goals, clear agendas, meeting norms and protocols, our district's leadership teams were and are able to work interdependently to achieve common goals, and function effectively. Celebration of successes comes with a price. Overcoming the frustrations of resistant, older colleagues, balancing family with all of our educational roles, and feeling like we are perceived to have all of the answers have contributed to feelings of frustration, as I detail later in the chapter, in a leadership activity.

These feelings are not at all uncommon to teacher leaders and I was both relieved and comforted to read Overcoming the Obstacles to Leadership, *Educational Leadership*, September, 2007. In it, Johnson and Donaldson captured the very essence of my frustrations and helped me to realize that I was and am not alone! The authors posited barriers as a triple threat. Teacher leaders' efforts are often blocked by cultural school norms of autonomy, egalitarianism, and deference to seniority.

One view that resistant teachers have is that teacher leaders intrude on their instructional space. They do not want someone coming into their room, observing, or offering feedback on their teaching. I see this barrier as perpetuating what may be ineffective instruction in that if the resistant teachers are not open to feedback to address any weaknesses in their teaching, then certain aspects of their teaching may remain ineffective. In addition, they may be closing off the rest of the staff to what may be working very well in their classrooms, thereby not allowing others to learn from them to improve their own practices. They perceive observation or feedback as evaluation, which is an administrator's role, not the role of the teacher leader. As more teachers in our district come to understand and embrace reflective practice and become mentor teachers, I see this threat diminishing.

Ensuring egalitarianism is another reason why resistant teachers may feel threatened. In my experience, resistant teachers' perceptions that the teacher leader has an attitude of I am more of the expert here is very real in schools today. Unfortunately, teacher leaders with this attitude do exist, even in my district. Resistant teachers question the possibility that one of their peers could actually have expert knowledge or even insight to share. In teaching my graduate level course, I have had firsthand experience with teachers who have felt threatened in this way. This feeling represents an inequality issue that resistant colleagues project onto teacher leaders unjustly, often times setting them up to fail. I have found that by asking and listening, building on what resistant colleagues express as needs rather than presupposing that I already know what is needed, and openness to learning from them, valuing their expertise, and admitting to not having all of the answers, all help to alleviate this threat.

A third perceived threat of teacher leaders, according to Johnson and Donaldson (2007), was deference to seniority reflected in the resistant teachers' attitude of, who does she think she is, after all, I've been teaching longer than she's been alive. Seniority to them equates to years in the profession. Some of our up and coming teacher leaders are, in fact, much younger than the veteran teachers, but not all. Nevertheless, they frequently face this threat. At times it appears to be a lose–lose situation; the resisters resist, and the teacher leaders give up. One young teacher leader is becoming so tired of the fight with her more senior grade level team-mates that she is close to resigning from her leadership role on the team. In order to keep from admitting defeat, how does a teacher leader, especially a young one, overcome these obstacles and become an effective teacher leader?

Each time I encounter the many definitions of leadership (Danielson, 2007; Dozier, 2007; DuFour et al., 2006; Gabriel, 2005; Lambert, 2003; Lipton & Wellman, 2007; Lipton, Wellman, & Humbard, 2003, Marzano et al., 2005; Moore-Johnson & Donaldson, 2007), I extract bits and pieces to create my own meaning of teacher leadership. My definition of teacher leadership centers on shared leadership – sharing knowledge with others, learning with and from others, and ultimately mentoring in this shared leadership approach, rather than mentor = expert, new teacher = student. First and foremost, I believe that teacher leadership must be promoted and encouraged. Teacher leadership cannot flourish strictly as a grassroots movement in the absence of administrative support. Administrators must acknowledge that everyone has a voice and that the voices need to be heard. Neither can they dictate that leadership happen in the absence of teachers who want to lead. Charlotte Danielson (2007) offered four conditions that promote teacher leadership.

1. School administration must provide a safe environment for risk taking. For many teacher leaders, this is new and is considered a risk.
2. Administrators must encourage teacher leaders.

3. Administrators must provide opportunities to learn leadership skills.
4. Administrators need to be aware of what Danielson refers to as the "tall poppy syndrome," a term in Australia referring to "those who stick their heads up risk being cut down to size" (2007, p. 19).

In our district these conditions are interwoven. Those of us who are encouraged into teacher leader roles stick our heads up, and feel safe doing so because we know we will be given opportunities to learn the skills we need. After all, what good is it to encourage teacher leadership without providing a safe environment or opportunities to learn the skills necessary to be effective? Only recently in our district have teacher leaders been presented with streamlined, differentiated professional development based on the need to gain the leadership skills essential for us to be effective. Our administration is now offering well thought out, meaningful, professional development in the realm of leadership based on having asked teacher leaders what their needs were in order for them to become more effective in their roles. Asking them created ownership and showed them that their needs were important. Once the needs were established, our Director of Special Services conveyed them to a professional developer (who had previously worked with all district administrators) and incorporated them into a class offered for graduate level credit. This class is being taught, and although the group is small, the content is rich. Some of us in the class, including me, are realizing that we need to learn more about our own strengths and weaknesses within our leadership styles, communication preferences, and personality traits. One of the most practical suggestions I have gained from the leadership class comes from Rosamund Stone Zander and Benjamin Zander's *The Art of Possibility* (2000). The authors offered us Rule Number 6, "Don't take yourself so g_damn seriously" (p. 79); "...lighten up, which may well light up those around you" (p. 80). I have found it to work remarkably well! Not only in teaching my graduate level classes, but in my role as RtI/ Mentor Coordinator, I usually exhibit a sense of humor and have been known to laugh quite often, as it is in my nature to do so. I may even be laughing at myself, but if it lightens up those around me and leads to other chuckles or smiles, our time together is not only fun but meaningful in a relaxed atmosphere.

Leadership Activity

One of my beliefs in shared leadership is that everyone has a voice and that voice is to be heard. After all, this is what brings us to shared knowledge. As a teacher leader facilitating various leadership teams, some successful and some not, I have found myself becoming a sounding board for other frustrated teacher leaders, who have come to trust me, in private settings. Often these situations were their opportunity for venting, knowing that I would listen. At other times, they needed to share frustrations that required an action to be taken. Regardless, these experiences prompted me to create my leadership activity for this chapter. After all, not only was I feeling similar frustrations, I wanted to be more than a sounding board. Although I wanted to honor that their voices were heard, by active listening, I wanted to take it beyond the literal interpretation to action. If I could let their voices be heard so that I and others may learn from their experiences, their venting could be meaningful and productive.

I have designed a Leadership Activity in which a teacher leader or an administrator could ask members of various leadership teams to reflect upon their leadership journeys. Although I explain the leadership activity in the first person, readers can generalize this to an activity which they too can do. Specifically, I asked,

- What does it take to be a successful leader in education?
- What has been your biggest obstacle and how have you managed to overcome it?

One could easily create other questions, depending upon the information to be gathered. I quickly decided upon email as the mode for the questionnaire, which I thoughtfully composed and sent out to all members of four different leadership teams. It is important to note that I prefaced the questions by asking them for the favor of their response. Responding was not required, but by letting them know that I valued their feedback and considered them to be successful educational leaders in our district, I had hoped this would increase the number of responses I received. Also, I was strategic in the timing. I knew the beginning of the school year was hectic enough, and that these teachers had enough on their plates. I acknowledged this in the email, thereby validating their time constraints and busy schedules. I gave the teachers a few weeks to settle in before sending out the email, and then a week to respond. Finally, the writing style with which I am most comfortable leans toward informal and friendly, with a bit of humor thrown in. Clearly, this was my email and having fairly healthy relationships with all of the recipients since I had taught with these teachers, some of whom were close friends, for years, I felt confident that I would hear back from many of them.

I chose email format for a few different reasons. Email allowed me to save time in both distributing the questions and gathering their answers. I had hoped that written reflections would also allow for the feedback to be more honest. Often it can be easier to be more truthful behind the safety of a computer screen than in a face to face interaction. I also assumed that teachers would be more detailed in their reflections, eliminating some of the sidebars and slang of casual conversations.

Since feelings of frustration are not limited to teachers in our own small district, but rather are common to many other districts, this leadership activity is something that teacher leaders can do in their local areas and can help to create a feeling of support for new learning to occur. Similarly, other districts may utilize this same activity, changing the questions to ones that pertain to their own cultural issues. Doing this activity as a teacher leader cohort, listening, and including everyone shows that teacher leaders' ideas and voices are heard and valued. The responses from many teacher leaders in my district from having engaged in this leadership activity comprise the Leadership Response section of my chapter in the hope to offer suggestions to support others in similar situations.

Leadership Response

Although I received many responses, I have chosen to highlight three teacher leaders' voices for different reasons. The first reason is that the three teacher leaders are actively involved in two different leadership teams. Second, they generally wrote the most detailed responses. Third, their responses represent different grade levels and teaching areas and offer different beliefs as to their obstacles and what it takes to be successful teacher leaders. Fourth, they represent a compilation of common themes in all of the responses.

The first response is from a high school English teacher who has been teaching for 15 years in our district. She is also a mother of two teenagers, and a coach. The second response is from a kindergarten teacher who taught in a private school setting for seven years before coming to our district a year and a half ago. The third response is from a teacher of English Language Learners at the elementary level, who had taught previously in our district, moved away to teach for five years and then returned last year. Collectively, their responses include many of the obstacles and suggestions offered by the entire group.

The veteran high school English teacher struggles with colleagues who are "filled with animosity and resentment," but tagged her own impatience with colleagues who fight change as her biggest obstacle. She states, "My impatience blinds me to empathizing with others who need time to process and accept innovations and improvements." Just this awareness is contributing to her becoming a better leader. She credits motivation, fed by a passion for student learning, as key to successful leadership. Motivation is essential in finding balance in roles to students and colleagues. What gives her the motivation to continue in her leadership roles? "The students are the reason I am here and why I love teaching so much. . .through my leadership, I am able to touch all kids at all grade levels by working with my colleagues." Additionally, she admits that working and learning from her colleagues has helped her to become a better teacher.

The Kindergarten teacher new to the district a year and a half ago notes that the biggest obstacle in her teacher leadership journey and what it takes to be a successful teacher leader are one in the same! "Being a leader in education does not mean you have all the answers, it is just a willingness to increase learning." She also writes that this is also frustrating for her because others perceive her as having all answers. She realizes that this perception is not true. She believes, "It is not about having all the answers, it's about the willingness to communicate and work with others to find the best possible solution." This is not always easy, and thankfully, she acknowledges that, "Being a strong leader is knowing when to ask for help."

Our English Language Learner teacher, who also happens to be a bus driver and after school Spanish teacher, offers a willingness to take risks, learning from others, sharing what you know, flexibility, and a positive attitude as keys to successful leadership. Her positive attitude has gotten her through the biggest obstacles, which she states "are often the little things, specifically, things that are opposite of what it takes to be successful!" She is so positive; she even stated the obstacles in a positive way! To translate, I can only assume she means unwillingness to takes risks, unwillingness to learn from others, unwillingness to share what you know, not being flexible, and a negative attitude as obstacles. "Leaders are people who have held onto dreams and who help others to build and hold onto their dreams too."

In summary, how does one help others hold onto dreams? Since I value honoring each voice, in addition to the responses above, I have also included quotes from other teacher leaders who I queried and who offered additional insights. To respect the teachers' wishes, their identities will not be disclosed in the reference section. Here is a listing of others' valuable thoughts on what it takes to be a successful teacher leader.

"Leaders are honest with themselves and others about their roles."
"Leaders possess honesty and integrity."
"Leaders are willing to see all sides of an issue and not pass judgment."
"Leaders question their own assumptions and positions in order to see the focus or goal more clearly."
"Leaders know when to stop and take a look at something because it may not be working well."
"Leaders are willing to learn from others and share what they know."
"Leaders are willing to take a risk, when others are not."
"Leaders look forward to change and get excited about what's around the next bend."
"Leaders have colleagues who work together towards the same goal."
"Leaders are people who care more for others than for themselves."
"Leaders strive to gain new insights which involves a willingness and openness to share and accept others insights."

I honor the courage it takes to be a teacher leader today.

References

Colorado Department of Education (2007). *RtI in Colorado: Response to instruction & intervention.* Retrieved August 13, 2007, from http://www.cde.state.co.us/cdesped/download/pdf/RTI_DefPrinciples_July2007.pdf

Danielson, C. (2007). The many faces of leadership. *Educational Leadership, 65*, 14–19.

Darling-Hammond, L. (1997). *The right to learn.* San Francisco: Jossey-Bass.

Dozier, T. (2007). Turning good teachers into great leaders. *Educational Leadership, 65*, 54–58.

DuFour, R., DuFour, R., Eaker, R., & Many, T. (2006). *Learning by doing: A handbook for professional learning communities at work.* Bloomington, IN: Solution Tree.

Gabriel, J. (2005). *How to thrive as a teacher leader.* Alexandria, VA: Association for Supervision and Curriculum Development.

Johnson, S. M., & Donaldson, M. L. (2007). Overcoming the obstacles to leadership. *Educational Leadership, 65*, 8–13.

Lambert, L. (2003). *Leadership capacity for lasting school improvement.* Alexandria, VA: Association for Supervision and Curriculum Development.

Lipton, L., Wellman, B., & Humbard, C. (2003). *Mentoring matters: A practical guide to learning-focused relationships.* Sherman, CT: MiraVia, LLC.

Lipton, L., & Wellman, B. (2007). How to talk so teachers listen. *Educational Leadership, 65*, 30–34.

Marzano, R. J., Pickering, D. J., & Pollock, J. E. (2001). *Classroom instruction that works: Research-based strategies for increasing student achievement*. Alexandria, VA: Association for Supervision and Curriculum Development.

Marzano, R. J., Waters, T., & McNulty, B. A. (2005). *School leadership that works: From research to results*. Alexandria, VA: Association for Supervision and Curriculum Development.

Pollock, J. E. (2007). *Improving student learning one teacher at a time*. Alexandria, VA: Association for Supervision and Curriculum Development.

Zander, R., & Zander, B. (2000). *The art of possibility*. New York: Penguin Group.

Part V
New Seeds: Hopes and Visions for the Future

Chapter 14
A Beginning Teacher's Reflection: I'm a Teacher, but Is Teaching Alone Good Enough for Me?

Wendy Files

My Story

I'm not sure I'm an educator. I'm a teacher, I'm a scientist, I'm a wife, and I'm a twenty-something. However, do these things together turn me into an educator? My own education seems only to have just begun. I attend conferences with WISE women of science and see only glimpses of myself in their rarified art. How do I get there from here? Beyond not seeing clearly into the development of my own scientist-educator future, I seem to be asking an additional question that is worth answering first: How did I come here? That is my uncertainty.

I found a dinosaur bone once. It was obviously the leg bone of a dinosaur that had lived in the cow pasture behind my childhood home. I was so excited that I took the bone up to the house to show it off. Unfortunately, my parents were not as excited, as they shared that it was most likely a bone from the cow we had buried last summer. Disappointed, but not discouraged, my interest in science continued to develop. As I grew older, I found myself spending less time digging around in the cow pasture and more time pursuing science through avenues traditionally available to a young student.

When it was time to head off to college, I chose to study biology. During my first two years at school, I spent much of my free time in the practice of science. I studied salmon in the Pacific Northwest, I worked in the labs on the various campuses that I attended, and I did some part time work as a dental assistant. I loved all the work that I was doing, and I was enjoying my classes but all too soon my college experience was coming to an end. No one career path was more appealing than any other. Very concerned about what I was going to do after graduation, I appealed to both my parents and counselors on campus. We discovered an ongoing trend in what I enjoyed about my jobs in the field. When I worked in the labs, I had loved the hard science of researching. While working on the river however, my favorite part of each week had been when the local middle school students would

W. Files (✉)
Baker Middle School, Baker City, OR, USA
e-mail: wendyfiles@hotmail.com

K.C. Wieseman, M.H. Weinburgh (eds.), *Women's Experiences in Leadership in K-16 Science Education Communities, Becoming and Being*, DOI 10.1007/978-90-481-2239-4_14,

come out to the fish farm to help with the sampling and tagging. The realization that I enjoyed the people with whom I was working is what led me to the education department.

This realization was particularly difficult since I have held the belief for a long time that education was/is a cop-out. If I were good enough in science, I would be in science, not in education. This is a conflict I have had and with which I still frequently struggle.

Because of this conflict, it was a difficult transition for me to go from the straightforward, lecture-exam structure of the science department to the more student centered approach in education. I was amazed when professors suggested we draw posters to show the relationships between teacher and student and then share them with our small group. What happened to the reassuring task of memorizing several hundred terms and taking an exam? Although the entire transition seemed difficult, the most challenging aspect was the introduction of daily written assignments. I had very wisely chosen the sciences knowing that I would have few written expectations. Fewer written expectations were preferable because I have always been and still am afraid of written communication. This fear of writing arose from my experience with a high school English teacher who told me that my writing was never going to be any good. This comment has greatly shaped and limited me as a student and teacher. As a result, in college I chose to take as few writing courses as possible, fearing that I would fail, and to this day I feel unsettled to have anyone read my writing.

Furthermore, I had never considered education as a career option prior to this. Growing up I had not had any particularly great experiences at school, and I was not even convinced that my teachers were any good. Was I settling for what seemed like a sure career path that did not involve immediately going to graduate school? Would I be any better than the people who had so crudely guided my own young learning experiences? Thus, I plodded into my education coursework with guarded enthusiasm and before I knew it, I was pomp and circumstancing my way into the gymnasium to receive a B.A. in Biology with an Education minor and eligibility for a secondary sciences teaching license.

My first school year as a teacher was filled with excitement. I am sure the school district in which I was teaching did not carry enough liability insurance for those first few labs. Despite being terrified at least half of the time, I was truly enjoying myself. I loved sharing science with my students, and I loved being on my own in a classroom that was mine. I was beginning to anticipate the challenges, stress and rewards of classroom teaching and was spending less time trying to keep my feet under me.

Beyond the typical challenges within a classroom, I was also dealing with the additional obstacles of becoming licensed to teach in Oregon. My original license was in Colorado, and to my dismay, Oregon would not fully accept it. Instead, I was given three years to complete the additional requirements set forth by the state of Oregon, which included two basic knowledge exams, one general subject area exam as well as at least seven PRAXIS exams that covered material from each area of science.

Discouraged and nervous, I spent much of my first two years of teaching panicking over, studying for, and creatively financing these exams. Also discouraging was knowing that with my husband's career in aviation, it was very likely that we would move again, probably out of state. Unfortunately, relocation would mean I would once again have to prove myself as a capable teacher.

Luckily, by my third year the licensure exams were tapering off. As a result, I decided to go back to school for a master's degree. I chose a program in secondary science education with Western Governors University because of its focus in the sciences. It turned out to be a wonderful program as well as my best year teaching yet. During that year I was able to teach, (which I was learning to truly enjoy), and also study science. It was the best of both worlds. The most eye-opening aspect of my master's degree was the research-based thesis project I completed at the end of the program. For the first time, I used my interest in research to attempt to study and improve my classroom. This was perfect for me. Not only was I teaching, but I was researching about my science teaching while teaching!

Finally, as anticipated, my fourth year of teaching was accompanied by a move from rural Oregon to Albuquerque, New Mexico. When my husband and I knew that we were moving, I applied for many science teaching jobs and had hoped to stay with middle school science so that I could continue my previous research. To my shock however, I was offered a full-time chemistry position at a well-respected high school just outside the city. Knowing very well that this was my weakest content area, I was very scared. However, I also knew that accepting the job would be a great opportunity to learn and become a better teacher. So, I accepted the challenge and as a result learned more content and safety strategies than I ever imagined. During this year I spent less time researching my teaching techniques and more time watching and learning from others.

At the end of this year we moved again, putting my next (and fifth) year of teaching back in Oregon, again in middle school sciences. This job will hopefully allow me to continue the middle school research that I had loved and left behind only a year previous. In the search for my niche within education I am participating in as many activities as possible within the field. Beyond my current day job, I am also tutoring online, working on research within my classroom, and teaching summer school when possible. I am hoping that in the future I can somehow use my teaching skills in an online learning environment to meet my professional needs as an educator, and my personal needs of living in a rural location and spending time with my family.

During my first three years of teaching I repeatedly asked myself why I kept coming back to the classroom. During those years I was the only science teacher in my rural district and finding a mentor was very difficult. However, I was able to travel to another district and observe other science teachers several times, from whom I learned much. It was not until my fourth year of teaching in an urban district that I truly experienced an ongoing and, ultimately, very rewarding mentorship. During this year I was assigned to work very closely with another chemistry teacher. We built our lessons together, shared our classroom concerns daily, shared common assessments as well as assessment results and were openly and easily able to observe each other teaching when desired. From this experience I believe that I was able to improve my teaching in that one year far more than I had in the three years that I was teaching alone at my first job.

My current school, again in a rural district, does not have an official mentoring program but I intend to continue to use the mentoring skills that I have learned. Toward this end, I and the other two teachers at my school are working together to build a professional learning community in which we meet weekly to share information for bettering our classroom learning environments. It is through these types of collaborative experiences that we become better educators.

I look at where I am now as a teacher; it is easier now than when I began teaching to see where I am going. I face questions that I sense may be universal to women educators, particularly in science: How does my desire to be involved in laboratory/research science fit within the framework of my teaching career? And is teaching really enough for me? What mentoring do I need to be the best teacher I can be?

From discussions with my advisor from graduate school I understand that a Ph.D. program could be a next step in my career. I do not know if I am ready. I do not know if a Ph.D. program is what I want to do. Now that I may have found a way to incorporate my need for research into bettering both myself as a teacher and my classroom, do I want to do something else? Can I be happy continuing to teach secondary school students?

So although I have glimpses of where I might want to go, I do not have any solid answers yet, and I may never have them all. . .but I do know that I can finally admit that I love being an educator. I simply do not yet know what type and at what level.

Theoretical Framework Intertwined with Aspects of My Story

The nagging questions that I have presented in my story focus around three main themes, self-efficacy, mentoring, and professional learning communities for mutual mentoring. The definitions that I will be examining and using for the three themes are as follows:

- Self-efficacy: "people's beliefs about their capabilities to produce designated levels of performance that exercise influence over events that affect their lives" (Bandura, 1994, para. 1).
- Mentoring: "Simply stated, mentoring is a process in which a more experienced person supports and aids a less experienced person in his/her professional or personal development" (Georgia Department of Technical and Adult Education, 2007, para. 1).
- Professional learning communities for mutual mentoring: "a group of educators committed to working collaboratively in ongoing processes of collective inquiry and action research in order to achieve better results for the students they serve" (DuFour & DuFour, n.d., p. 3, para. 3).

I believe that the above themes greatly affect who I am as a teacher as well as whom I am to become.

Self-Efficacy

According to social cognitive theory, "individuals possess self-beliefs that enable them to exercise a measure of control over their thoughts, feelings, and actions" (Pajares, 2002). This belief system is highly influenced by how we view our previous accomplishments and capabilities. These ideas also suggest that although we have some control over the outcomes of our lives, much of what we do is influenced by our unconscious mind.

According to Bandura (1994), a person's strong self-efficacy enhances human accomplishments while poor self-efficacy may lead to low aspirations and weak commitment to goals. He believed that there were four main factors that could help improve one's self-efficacy. These factors included experiencing success or overcoming hardships, observing others similar to oneself overcome or achieve things, experiencing social persuasion by peers that one is capable, and lastly, having an emotional awareness of one's capabilities (Bandura, 1994).

These prevailing thoughts in turn not only influence us, but also those around us. This may be of particular interest to us as educators since research shows that our self-efficacy can affect our classrooms and in turn our students (Pajares, 2002). Furthermore, Pajares stated that educators may have an effect on students' academic performance and achievement in school as well as what majors and careers they may choose.

My own self-efficacy has been greatly impacted by the educators around me. My previous experiences with educators as well as my own experiences in education are some of the things that influence how I see myself and how I am willing to accept or deny challenges set forth. So far, my largest challenge has been simply seeing myself as an educator. Because my previous interactions were not particularly positive in education, I needed (and need) to overcome my misconceptions of educators so that I could (and can) accept myself as one. I also needed (and need) to learn to see myself as a role-model or a mentor to students so that I can accept the responsibilities that come along with being an educator.

I wonder if it was my low self-efficacy that led to me to the path of education in the first place. Because growing up I did not have particularly good experiences within the classroom, nor did I view my teachers as highly qualified, I have wondered if I ended up in education because I did not think I could do any better. If I had thought of myself more highly, would I have tried to further myself in the hard sciences and attended graduate school directly after college? I do not know. Looking back I do not think it would have made me any happier than I am now, but I can also see that it was fear of failing that kept me from pursuing the hard sciences. Knowing what I know now about my own self-efficacy motivates me to reach students like myself and to let them know that they can succeed, even if they do not believe it themselves quite yet.

These conceptions are held deeply within me and although I am slowly seeing them change and less drive how I see myself, I still constantly fight with them. Through this personal struggle I am attempting to challenge not only what I believe about myself and what I am capable of doing, but also I am trying to use my previous experiences in education to make my teaching better. My challenge is to better meet the needs of my students, to ensure that my students do not view educators in the same light that I did and to allow myself to fully accept and take pride in the title of being an educator.

This is a learning process for me, as are many things in life. As an educator you quickly learn that although you may have finished your undergraduate degree and maybe even another degree, you are never truly done. "Lifelong learning is more than just education and training beyond formal schooling. A lifelong learning framework encompasses learning throughout the life cycle, from birth to grave and in different learning environments, formal, non-formal and informal" (The World Bank Group, 2001, para. 1). I learn new methods to teach and materials to use, and more often than not, I am learning in order to make myself a better educator.

From what I have learned so far, becoming a true educator is a learning task that is never quite finished. As each year I teach passes, I am learning more and more about myself as a teacher and also about the teacher that I want to be. To me, being an educator means learning to master being a student as well. By this I mean that in order to improve myself as a teacher, I always need to be willing to learn how to improve. The best way that I have found to improve myself as an educator is through self evaluation and mentoring.

Mentoring

According to reports released by the Alliance for Excellent Education (2004), 14 percent of first time teachers quit in the first year; within three years 33 percent will leave. These statistics may seem startling, but additional research shows that successful mentoring can increase the retention among first year teachers. In the early 1990s the University of Montana conducted an experiment to see if there was a relationship between mentoring programs and teacher retention. They found that after the first year "92% of the mentored teachers continued teaching, compared to 73% of the non-mentored" (Ingersoll & Kralik, 2004, para. 54) that continued teaching. Similar results were found in California, New York City, and Toronto. At the time of my first year teaching, I was unaware of these statistics. I felt the pressures that I am sure many first year teachers feel, which at times did and often still have me wondering what keeps me coming back to teaching.

Mentoring is an act of an experienced person helping a less experienced individual. The Virginia Board of Education [VBE] (2000) stated that "Mentor programs help beginning teachers make a successful transition into teaching by relying on the expertise of veterans to provide a clinical, real-world training process" (p. 6). The VBE elaborated on the activities that should be priorities of any teacher mentoring program. These priorities include an open communication between the mentor and mentee, time during the scheduled work-day to meet, strategic placement or additional time if the teacher is given a difficult teaching placement, and ongoing professional development activities that are responsive to the teacher's needs (VBE, 2000). These are all great foundations for a successful mentoring program and hopefully will increase teacher retention (VBE, 2000).

Mentoring to me is simply the act of observing and collaborating with other experienced teachers or educators. Ideally, mentoring is not a one-time event, but instead a steady and ongoing open arena in which a mentee and a mentor share teaching ideas, collaborate on projects, and improve teaching strategies.

Because I am relatively new to education, most of my experiences have been as the mentee, and I have experienced both weak and strong mentoring programs. My first year of teaching was a year without formal mentoring assistance. Because my school did not have any other science teachers or new teachers, they did not have an established mentoring program. I was left to my own devices. Luckily, I found another teacher who was willing to work with me and act as my informal mentor. We shared with each other our common concerns about students, grades, and curriculum. This teacher also walked me through my first few labs of the year. These conversations and snippets of advice were the only mentoring I received. Thankfully, I did not know what I was missing.

It was not until my fourth year of teaching that I moved to an urban district that had an established mentoring program. However, the program was set up only for teachers new to the profession, and since I was no longer considered new, I was not a part of the official mentoring program. I was assigned a fellow teacher, however, who worked very closely with me throughout the year. This mentoring experience included classroom observations during which my mentoring teacher would watch me and other times when I would watch her. We would then share with one another, making sure to include both positive and negative things about what we had observed. We also created lesson plans together and common assessments, all of which helped reduce the burden on myself from being new to the district and showed me many new ways to improve my teaching.

My teaching was improved in many aspects. First, my understanding of chemistry became much deeper with the help of my mentor who had more traditional schooling in chemistry than I did. My mentor was also a much more experienced teacher than I was. She was able to provide me with ideas on how to make science material easier for students to understand, how to better organize my classroom, and even how to better organize my class work. All these ideas helped me to be a better teacher.

Professional Learning Communities for Mutual Mentoring

In *The Power of Professional Learning Communities*, DuFour and DuFour (n.d.) defined a professional learning community (PLC) as: "a group of educators committed to working collaboratively in ongoing processes of collective inquiry and action research in order to achieve better results for the students they serve (DuFour & DuFour, June 12, p. 3 para. 3)." These PLCs or groups meet regularly, have common curricular goals that are aligned with state assessment guides and attempt to teach them at a common pace (Schmoker, 2006). To accomplish these goals the PLC spends their time in their regular meetings focusing on concrete terms of each lesson as well as on ways to assess the success of the lessons taught (Schmoker, 2006). An analysis of the results leads to improvements in instruction (Schmoker, 2006).

A unique aspect of this urban district in which I worked was the presence of a strong professional learning community within the chemistry department. Our PLC consisted of about 5–7 chemistry teachers who met on a weekly basis. We discussed success and failures within the classroom, common concerns with the material, assessments, and other forms of advice for one another to apply within the classroom. The combination of the individual mentor and team PLC provided me with a very strong mentoring experience.

Since my new school district does not have an established PLC among the three science teachers, we have spent much of this year creating our own. Compared to the PLC of which I was a part in New Mexico, this PLC is very informal. The two other science teachers and I try to meet at least once a month. During these informal meetings we discuss everything from what we have recently taught and what we are going to teach to what is working and what we have found is not working within the classroom. These meetings although simple help each of us to better meet the needs of our students through collaboration and communication.

Leadership Activity

Because I am still relatively new to education it is easy to see myself as a team member or a teacher in need of mentoring, and it is far more difficult to see myself as a leader. I think that only now, after five years of teaching am I beginning to feel confident in myself as a leader in my own classroom. Seeing myself as a leader in my field is still very difficult because I feel as though I still need so much leading. I often find myself acting/playing the role of a leader in my job each day as a teacher, yet I am also still searching for leadership in my future. I am constantly asking myself two questions: Where am I going? Who is my leader/mentor? In response to these questions which may be questions that other teachers early in their careers ask, I can respond with several important suggestions.

(1) Ask yourself about your priorities in life as the answers to these questions define what future leadership looks like for you. The questions I have asked about my priorities are:

- What do I want to be when I grow up?
- Does it meet my needs to be with my husband, to one day have kids?

- Does it mean that I will be in school forever?
- Will I ever be able to relax?
- Do I need to know right now?

(2) Continue to search for leaders/mentors to guide your teaching.
(3) Continue to research options within a teaching career.
(4) Work to make as many contacts as possible in search of what peaks your interests including both educators and noneducators in your area. For me this has included previous college instructors, teachers within my district, and in other districts both in my content area as well as outside, and also professionals in other various fields of science.

The process of becoming a leader is a mental activity within as well as a social and physical activity. It is about experiencing life in order to see which priorities are really the most important.

Strategy for Mentoring to Become a Leader in the Classroom

Like many others who have come before me and advised fellow new teachers, I would advise the following.

(1) Look for good mentors, people who have experience with both the content as well as time within the classroom and who are willing to collaborate and share both successful and unsuccessful experiences with you.
(2) Be open to being mentored.
(3) Find a mentor with whom you are compatible (personalities, working style, expectations, etc.) and can work well. Questions that a new teacher can use to find a good mentor are:

- How much experience does the teacher have? Is this amount of experience enough for me?
- Does the teacher really know the subject area?
- How does the teacher think and approach tasks or responsibilities as a teacher?
- How do I (young/new teacher) think and approach tasks and responsibilities as a teacher?
- Does the teacher's answers match mine (work for me)? Is the teacher like me, and do I need the teacher to be like me for the match to work?

(4) Find time to observe and learn from your mentor teacher, and in return be open to having the mentor teacher come, observe, and reflect with you about your classroom. To truly improve your classroom teaching, you need to be willing to let others provide you with feedback on your teaching.

Developing an Informal Professional Learning Community

Establishing a PLC can be difficult depending on the willingness of the individuals within your department, as well as the willingness of your school or district. In order to begin however, I recommend the following:

(1) Meet with the members of your department and review with them the benefits of a strong PLC to both them and their students.
(2) Find a meeting time that each member of your team is available to collaborate.
(3) Meet regularly and keep meetings short and to the point. Focus on sharing and improving teaching strategies, techniques, and developing common curricular goals and assessments.
(4) Share the results from common assessments and use the results to initiate discussion on how to best present classroom information.
(5) Do not use assessment results to find fault in fellow educators but focus simply on improving everyone's teaching in order to best meet the needs of the students.

Leadership Response

I have found, and am still finding, that the suggestions I made in the previous section are very helpful. I have also found that the act of articulating my ideas about mentoring has led to professional and personal growth. The acts of being aware of yourself and listening to your inner concerns, questions and desires help you to better understand yourself. Through personal reflection you learn about yourself, and through a good mentoring relationship you can learn to better fit into your role as an educator.

Based on a sense that growing into being a successful teacher is a process fraught with at least psychological peril, we are left with a challenge to guide our newest colleagues as they negotiate their first experiences. What can we do as educators to help first year teachers reach their potential? What can we do to make sure that new teachers are teaching the curriculum to fidelity? How can we use our experience and knowledge to make the teaching process easier, better, more efficient for the educators around us? These are questions that have little to do specifically with science education but are broad enough in their scope to include any content or specialty areas.

As a science teacher myself, I can only directly address the ways that we can support beginning science teachers. Two issues that are worth mentioning confronted me early in my teaching career. First, as a student of science I often walked into the lab and found material and apparatus set out. When the activity was complete, I walked away and let someone else deal with disposal and waste management. No one helped me understand how to manage a lab supply room, dispose of waste materials in an environmentally safe way or deal with the hazards of teaching in a lab. As I mentioned earlier, few insurance companies would have knowingly underwritten my first few labs. Second, I received very little instruction on how to have appropriate expectations at various grade levels for material that I knew extensively from four years in college biology classes. How much genetics does a junior in high school really need to know? Moreover, how do I take my knowledge and thin-slice it into pieces that a high school student can grasp and still have that student come out with a reasonable base of knowledge? These are the questions that haunted my first year as a teacher.

As a teacher with more experience, I look back on these and other considerations of my early teaching career and see the immense value of developing self-awareness and having a support network of people who knew more than I did. My call to action is twofold. Mentors must be available to educators in a way that is useful without compromising the planning and preparation time of the teacher. Teachers must be encouraged to carefully reflect on their knowledge and abilities and to act accordingly.

In my own experience, gathering a professional learning community around oneself can be as formal as a district-mandated, time-allotted activity or as informal as a breakfast over coffee to make sure that there is some degree of continuity across the classes. While each has its associated benefits and perils, there do seem to be a few required elements in either case. Each of the ideas I propose originates directly from my own experiences in attempting to establish and maintain professional learning communities. First: a willingness to collaborate. In a recent meeting at the district level, I encountered a distinct interest from some of my peers outside the sciences to isolate their classrooms from the learning community that had developed to map the curriculum. This sense of "I intend to do what I've always done" has no place in the mentoring relationship. Second: a strong sense for the value of teaching fundamental skills. While important in science, this value is true for any content area. A professional learning community needs to be guided by a common purpose. Finally (and this is where self efficacy and mentoring intersect): the members of the mentoring relationship need to have an understanding of and a willingness to accept that the mentoring relationship is a two way street. For example, a first year teacher may have limited classroom management skills and may have fewer tools in their file cabinet but come prepared with a strong sense of the most current pedagogical avenues being explored in academic circles. The more experienced teacher may have strong management skills and a variety of pedagogical avenues that are not necessarily current. If each member accepts their dual role as a leader and a learner, the relationship is strengthened.

I have concluded that it is my own struggle to understand my role as a teacher, a leader, a parent, and a wife that leads me to look more deeply at the process of becoming a leader in the modern science education classroom. Perhaps I will find that my role in education remains at the secondary level. I do know that it is my responsibility to share my experiences with other new and beginning teachers, to help ease their transition into the field of education.

References

Alliance for Excellent Education (2004, November) Tapping the Potential: Retaining and developing high-quality new teacher. *Alliance for excellent education.* Retrieved September 21, 2007, from http://www.all4ed.org/publications/TappingThePotential/TappingThePotential.pdf#search='tapping%20the%20potential'

Bandura, A. (1994). Self-efficacy. In V. S. Ramachaudran (Ed.), *Encyclopedia of human behavior* (Vol. 4, pp. 71–81). New York: Academic Press. Retrieved January 14, 2008, from://www.des.emory.edu/mfp/BanEncy.html

DuFour, R. & DuFour, R. (n.d.). The power of professional learning communities. *The National Forum of Educational Administration and Supervision Journal.* Retrieved June 12, 2008, from http://www.allthingsplc.info/pdf/articles/ThePowerOfProfessionalLearningCommunitiesDuFours.pdf

Georgia Department of Technical and Adult Education. (n.d.). Mentoring program. *Georgia Department of Technical and Adult Education.* Retrieved December 14, 2007, from http://www.coe.uga.edu/chds/mentoring/mentoring_defined.htm.

Ingersoll, R., & Kralik, J. (2004, February). The impact of mentoring on teacher retention: What the research says. *Education Commission of the States.* Retrieved December 7, 2007, Website: http://www.ecs.org/clearinghouse/50/36/5036.htm

Pajares, F. (2002). *Overview of social cognitive theory and of self-efficacy.* Retrieved December 7, 2007, from http://www.emory.edu/EDUCATION/mfp/eff.html

Schmoker, M. (2006). *Results now.* Alexandria, VA: Association for Supervision and Curriculum Development.

The World Bank Group. (2001). Lifelong learning. *The World Bank Group.* Retrieved January 14, 2008, from http://www1.worldbank.org/education/lifelong_learning/overview.asp

Virginia Board of Education [VBE]. (2000). *Guidelines for mentor teacher programs for beginning and advanced teachers.* Retrieved June 12, 2008, from Division of Teacher Education and Licensure Web site: http://www.doe.virginia.gov/VDOE/newvdoe/legislat.PDF

Chapter 15
Teaching in a Rural Community, Where Roots Grow Strong and Risks are a Daily Occurrence

Sara Lamar

As you read my chapter, it will be useful to know that I am a first year teacher working part-time in a rural school district in Gunnison, Colorado. I recently completed an undergraduate secondary teaching licensure program at Western State College. I hold a B.A. in biology, and also have an interest in experiential education.

The Story of My Experience

Tree roots grow deep, strong, and widespread. They can anchor a 40-feet tall tree in 100 mile-an-hour winds, capture nutrients to sustain thousand-pound organisms, and live for hundreds of years. Humans can also cultivate roots. The roots stem from a variety of sources, often growing and digging in when one discovers home. Sometimes these roots can be left behind, only to be found again when one returns. Sometimes the roots get transplanted, such as in my case. I developed a few roots in my hometown in New England, but after moving have begun to nurture others where I currently reside. To some, roots are stronger in rural communities, and this can often create unforeseeable conflicts of interest. I love living in a small mountain community in Colorado, but have found it very difficult to find a job teaching secondary science. However, from the moment I first arrived in Gunnison, I have felt comfortable and safe. I came to this town as a sad and disheartened woman, yet quickly became a healthy, happy, and satisfied one. This is where I have healed and found myself. Although sacrifices have been made and desires have changed here, deep down inside I know that, right now, I could not live anywhere else.

I have lived in rural towns my entire life and am grateful for the opportunities this has given me. Living closer to nature has allowed me to develop my passions for hiking, skiing, mountain biking, swimming, mountain search and rescue, and science. Each season brings a new outdoor adventure in which to partake, a new scientific concept to study and appreciate. I have created a network of family, friends,

S. Lamar (✉)
Gunnison High School, Gunnison, CO, USA
e-mail: saradlamar@gmail.com

K.C. Wieseman, M.H. Weinburgh (eds.), *Women's Experiences in Leadership in K-16 Science Education Communities, Becoming and Being*, DOI 10.1007/978-90-481-2239-4_15,

and colleagues in the Gunnison Valley that I appreciate and trust, and I am not yet ready to try to build another one somewhere else. Regarding the unique scientific concepts to study, Gunnison is composed of a unique high altitude desert that I have yet to find anywhere else in the country. I am continuously impressed with the life that can grow here, and this love for biology has generated a passion for teaching others about the natural wonders of this world.

In the fall of 2002, I began my undergraduate education in Keene, New Hampshire, a town with a population of around 22,000, including the college students. Even this town was too big for me, and I dreamed of moving to a place where I was surrounded by mountains and other people who loved to enjoy them as I did. However, as much as I may have desired to leave Keene, it was certainly the place that first encouraged me to take risks. It was also where I discovered the wonders and challenges of experiential education. I owe this to an education professor who provided my class with the opportunity to design our own curriculum for the course, grade our own papers, and brainstorm ideas to make his teachings more effective and beneficial. At first, it was slightly uncomfortable to think outside the box and I struggled alongside my peers when creating assignments and goals for myself. Gradually, though, I fell in love with the challenge of being self-motivated and came to appreciate education on such a unique and simple level.

During my junior year at Keene State I participated in a national exchange program and moved to Gunnison for a semester at Western State College, where my best friend, Hannah was going to school. Hannah had moved to Gunnison when we all went off to college, and of all our successful friends, she was certainly the most pleased with her choice of an undergraduate program. We drove across the country together before the spring semester started, and after five days of visiting family and seeing the sights, we finally pulled into Gunnison. As exhausted as I was, I felt a strange and immediate connection with the town and knew that this was a place where I could be happy.

My roots now were transplanted and started to grow in Gunnison. I transferred to Western State College a few weeks after arriving in town. About a year later I walked with Hannah in our May 2006 graduation. I was also able to convince my younger brother to move west, and he fell in love with Gunnison just as hard as I had. During this time, I met my husband, Chip, who introduced me to skiing, laughed at my ridiculous jokes, and allowed me to be myself. Life was beginning to get busy, but I was not quite finished with school. I was to start student teaching the following fall at the Gunnison High School. I was excited to begin my student teaching and be that much closer to educating young minds in my own classroom.

During the eighteen or so weeks at the high school I taught nearly 100 freshmen the wonders of chemistry and taught environmental science to a small class of juniors and seniors. I was fortunate to work with two mentors who allowed and encouraged me to begin teaching from the first day, therefore getting to know my students and myself at a much deeper level.

At first, teaching was very challenging. I had some students with whom I could not connect no matter how hard I tried, and many of them stumbled over the difficult concepts that I taught. As much as I tried, I could not get them to appreciate their education as much as I did, even with experiential and self-directed means. However, each day I felt more and more comfortable in the classroom, and this attitude diffused to my students. Soon they were coming in during lunch and after school for help, and starting to take me seriously.

One day after a well-planned lesson that went exactly as I wanted, I fell in love with teaching. Never before had I imagined a feeling this intense. I knew that I wanted to spend the rest of my days educating students about the wonders of nature. The remainder of the semester flew by, and before I knew it, I was saying my goodbyes and finally receiving my college diploma. However, I did not have to say farewell to the high school just yet. During my time there, the registrar in the guidance office decided to retire and begin her own business. When the position was advertised in the local paper, I immediately jumped at the opportunity to stay in the high school near my students and wait for a teaching job. As there were no science teachers leaving the district in the near future, I accepted the registrar position in order to remain in the schools. Herein began my risky efforts to remain in the Gunnison Valley. One may wonder why staying is a risky behavior, when common knowledge may indicate that moving generates more of a risk. However, as I have stated before, my love for Gunnison has motivated me to live here, and therefore I must risk my financial and professional statuses to remain where I have built my home.

I worked in the guidance office at Gunnison High School for the remainder of the school year, and while I admired and respected my coworkers, everyday I longed to be back in the classroom. Students would approach me for help with their new science concepts, and I sat at my desk frustrated that I could not be the one teaching them this incredible knowledge. Every week I waited to hear about new openings in the schools; for a while they were all outside of my content area. However, I remained patient, and soon there was a job posted for which I could apply. One of my mentors during student teaching was taking a leave of absence for the fall semester, and her classroom was in need of a long-term substitute. I applied for this position and started wondering what I would do after Christmas break when I was no longer teaching, and someone else had been hired as the registrar. A few weeks later, another teaching job became available at the Crested Butte High School. Again, I took another risk applying for this job. Although Crested Butte is a part of the Gunnison Watershed District that includes Gunnison High School, I would be required to drive thirty minutes each way on a road that could become very dangerous during storms, which are a common occurrence during our long winters. I soon learned that I did not qualify for the Crested Butte position because they needed a combined math and science teacher. Due to Colorado law, I am not highly qualified to teach mathematics. Eventually, I accepted the part-time, semester-long teaching job at Gunnison High School for the teacher on leave. I teach in the mornings and work in the guidance center in the afternoons, content to be spending at least part of my day in a classroom.

I could have looked at other school districts, but as much as I miss teaching, I love living in Gunnison and cannot see myself moving just yet. Chip and I bought a house in Gunnison, were married in June of 2007, and developed a strong commitment to the Gunnison community. We are both active members of the local search and rescue team, have two dogs, and play in our neighboring surroundings as much as possible. Therefore, I am playing a waiting game right now, living each day without thinking ahead any farther than tomorrow. For me, living this way because I want to stay in Gunnison is riskier than planning and leaving this community.

Someone once told me that many people who do not live in rural areas are in disbelief when they hear that it is difficult to find a job in these communities. I suppose that I am an example of this situation. I have sacrificed my dream of teaching to work in any job that will pay the bills so that I may remain in a wilderness-oriented environment while I wait for that perfect job to show itself. One day I hope to make my desire to educate a reality, but until then, I am optimistic that I will find a way to be happy wherever I am due to the risks that I take in my life. Without taking chances, my life would be dull and monotonous. However, by taking risks, I am able to enhance my creativity, explore new places in Gunnison, and grow as both an individual and professional.

Theoretical Framework

There are several theoretical strings woven into my story. All, I believe, stem from my passion for experiential education, which is composed of two simple responsibilities: providing a direct experience with a learning environment for the student and facilitating a reflection on that particular experience (Joplin, 1995). First, how do I survive, and hopefully thrive, in a rural community where jobs are scarce and highly sought after? Secondly, is it healthy to take all the risks that I do everyday? And lastly, how do I promote these experiential practices into a classroom in a traditional public high school? With the self-motivation that I have cultivated since my freshman year of college, I will attempt to shed some light on these issues, and discuss the various theories that surround and support them.

I understand the importance of a sense of belonging and acceptance in a rural community, which then in turn enforces the need to take risks in order to survive in one. When I first moved to Gunnison, I knew only one person. I would go places and be the only individual present who did not recognize anyone else. However, it did not take me long to meet people in this friendly town, and soon I was being stopped for small talk in the grocery store. This quick, easy, and unconditional acceptance into the community that makes up the Gunnison Valley helped me to fall in love with the town. Now that I feel as if I belong here, I will do whatever it takes to stay. Maslow described belonging as a social need; humans need to feel acceptance in a community and need to love and be loved by others (Maslow, 1943). Community also brings about influence, connectedness, and a fulfillment of needs (Arcidiacono, Procentese, & Napoli, 2007).

I have lived in small towns my entire life, and it was within them that I developed my love for the natural sciences. Scents, sights, the slightest noise made by a cricket were all easier to sense in the quiet environment created by the woods around my house. Although I do not consider myself a feminist, Haraway's (1988) description of subjectivity toward feminism illustrates what both a rural community and nature are to me. She wrote that "the knowing self is partial in all its guises, never finished, whole, simply there and original; it is always constructed and stitched together imperfectly, and *therefore* able to join with another, to see together without claiming to be another" (p. 586). As I relate Haraway's quote to nature, I see her in an imperfect form, yet these unfinished, original characteristics make her whole and unique. The mystery surrounding our environment's way of life is what initially impassioned me to study science. And there really is no better place to love and appreciate the outdoors than in Gunnison, Colorado. To me, a small town is all about knowing and being comfortable with oneself in order to join neighbors who are just as imperfect, and help each other survive. With the support of my family and friends, I can confidently assert that I will endure, I will live through the challenges I face daily, and eventually, I will find a job that I love and with which I can find success.

Living in this small town has led me to take numerous risks, both personally and professionally, to get where I am today. Personally, I risk bodily injury while partaking in the activities I love, which include telemark skiing in the backcountry, mountain biking, hiking, mountaineering, and rescue. Professionally, I take risks to grow as an educator, such as trying innovative lesson plans, advocating for my beliefs in the classroom, and always doing what I feel is right. My age has also probably affected my risk taking behavior. Adolescence and early adulthood is a time for developing individual identity, opinions, and values (Miller 1989). However, as I grow older and continue to mature, I realize that I also tend to be a sensation-seeker. I feel "a need for varied, novel, and complex sensations and experiences and the willingness to take physical and social risks for the sake of such experiences" (Zuckerman, 1979, p. 11). As I understand the literature, the difference between risk taking and sensation-seeking is the degree to which one puts herself in harm's way. However, I also believe that many activities involve both risk and sensation-seeking. For example, in my personal life I love to backcountry ski. Every time a person goes into the backcountry, which I consider anything off a maintained road or a trailhead, they put themselves in harm's way, and therefore, are at risk. To clarify further, if I wanted to feel how deep the snow was on a certain slope, I could jump off a cliff and punch through the snow (sensation-seeking and risk taking), or I could simply push my ski pole into the snow until I felt the ground (risk taking). Using my pole to judge the depth of the snow still gives me the sensation of touch, and is far safer than jumping off a cliff. Because both instances take place in the backcountry, they both involve possible avalanche danger. They are both fairly risky, but the first is more likely to cause injury and to be more oriented toward sensation-seeking. As a teacher in a classroom setting, I could risk showing a violent movie so that my students really got to see the pains and tragedies of war (thus putting my career in jeopardy), or I could ask a war veteran to speak with my class about his or her experiences (a less upsetting situation, but still risky to a degree).

Taking risks has also enhanced my knowledge and education. "Learning is the reward of taking risks" (Young, 1991, p. 8). For instance, I have grown personally by taking chances outside my comfort zone as a leader on the local search and rescue team. One story that exemplifies this involves a woman who was lost in a nearby county in the fall of 2005. She was lost for four days and three nights, and I happened to be in the group that found her just as she was starting to give up hope. I was the most medically trained person there, and immediately took over caring for her health and injuries. At the time, this was significantly beyond my comfort zone, as she was the first person whose life was virtually in my hands as I helped rescue her. Hours later we extracted her from the backcountry and saw her off to a hospital. I still get email updates from her son; she seems to be doing very well and is enjoying her life like no one else can truly understand. As a leader of student learning in the classroom, I have learned several new teaching techniques by trying innovative lessons and have become more aware of, and added to, my strengths and areas for improvement. According to Young, I am an analytical risk taker, more inclined to scrutinize a situation and analyze important factors before participating in the task. I am "eager and willing to experience new learning and see this as its own reward" (p. 9). Young stated that this behavior is most common in the early childhood years, yet I am more inclined to believe that it is my gender that has caused me to continue taking risks. Wallach and Kogan (1964) conducted studies which revealed that a female subject, who was a risk taker, took the same degree of risks throughout her entire life, and a female subject who was not a risk taker took minimal risks throughout her life. Whatever my age or gender, many of the chances I take in my life are expected; my lifestyle choices ensure that I will be placed in harm's way more often than not. My love for the outdoors has led me to partake in incredible, yet somewhat dangerous adventures; nevertheless I plan to continue exploring new places and activities as often as possible. This attitude has in turn affected my professional life. A context of professional risk taking, which impacts my leadership development, is having the courage and patience to wait for my perfect teaching job to become available in my local area. Meanwhile several positive experiences are arising while I wait. From thinking a great deal about other ways through which I can pursue to impact our community's children, I decided to and am currently pursuing my master's degree in school counseling. I do not know where this decision will take me in my professional journey, but I am excited to see what possibilities will arise from continuing my own education.

Kahneman and Lovallo (1993) stated that "decision makers have a strong tendency to consider problems as unique" (p. 17). Leaders as decision makers can also tend to view problems as unique. Is that not how teachers are taught to think in classrooms as well? I was educated to believe that all students are different, and therefore their cognitive processes and learning styles are distinctive as well. Although I have never taught in an urban or suburban setting, I have heard enough stories to suppose that teachers in those schools face their own set of unique problems every day. And what an interesting statement about science – as researchers, professors, and lovers of scientific concepts, we look at scientific mysteries and queries as different, or else we certainly would not be a society as advanced as we are today. For example, take Holland's work (2007) in which he wrote about two students observing the mating dance of a black sicklebill, a large bird of paradise found in the mountain forests of New Guinea. After watching this amazing spectacle, one student might decide to investigate the history and rituals behind the complicated ballet performance, while the other may attempt to research the female bird's response to the male's attempt to reproduce. These two students watched the same spectacle, but have made different decisions about the problems they want to investigate. In making the decision to further examine the mating rituals of birds of paradise, they have become researchers, and therefore leaders. Researchers are leaders because they make the decision to investigate a problem to help society move forward. Research often brings about innovative ideas that could result in changes of thinking or behaving. Leaders also make decisions to help society move forward. Leadership activity brings about modern ideas and changes in thinking and behavior. I am a great deal like these students; I see myself as a researcher working to enhance my teaching philosophy. I observe other teachers and read literature, and then experiment with strategies that I like with my students as a teacher and counselor. Therefore, I am a leader in the classroom, working to bring about changes in my students' thinking and behavior to make them more effective and successful learners.

Experiential learning and experiential education have become cornerstones of my teaching philosophy. Joplin (1995) argued that all learning is experiential, and in order to comprehend, one must experience the concept by forming a significant connection. I think that experiential learning can be traced back to "My Pedagogic Creed" (Dewey, 1897), in which he stated that students should be taught to perform the fundamental activities found in society at the time. Dewey posited that schooling should promote the kind of expression and knowledge construction that aligned with these fundamental social and life activities. About 90 years later, Kolb (1984) developed an experiential learning theory in which he declared that learning was a process, not an outcome. In this process, drawing out students' beliefs and ideas about a topic, then examining, testing, and integrating these ideas into newer ideas was a way to relearn a concept. Kolb's theory defined learning as "the process whereby knowledge is created through the transformation of experience. Knowledge results from the combination of grasping and transforming experience" (p. 41).

Think of the student as an experiential learner and the teacher taking the role of an educator facilitating experiential learning. Each position has distinct roles but learning is the commonly shared process. Joplin's (1995) position was that experiential learning and experiential education go hand in hand. In experiential education the educator is responsible for providing a unique and direct encounter for the learner based on experiential learning, and then facilitating the following reflection of that experience. This reflection process, otherwise known as the action–reflection cycle (Joplin, 1995) continuously builds upon itself, growing deeper and more complex with each experience. An effective educator teaching within an experiential education perspective will nudge a student to begin this process, and then stand back and allow him/her to process their experience with minimal to no guidance from the teacher (Joplin, 1995). On the other hand, Druin, Owens, and Owen (1986) deduced that the student needs to be active in planning and carrying out activities, to assume responsibility for his/her own actions, and to interact with adults and peers. The teacher and student roles are often different from the ones played out in a traditional classroom, as students take on more and more of the teacher's responsibilities. After a student has led a day of hiking or planned out a community dinner and then reflected upon the experience, the teacher needs to give support and constructive feedback (Joplin, 1995). The ultimate goal is that the learner takes away a wealth of knowledge and confidence to a higher degree than he/she might have gained by more traditional education techniques (Joplin, 1995).

Not only do I believe experiential education to be the most effective way for me to educate my students, I also consider experiential education useful for the every-day learning experiences that occur outside the classroom. Many people wrongly assume that experiential education only encompasses ropes-courses, Outward Bound trips, and adventure-based events. However, this teaching approach is also useful in the more traditional classrooms found in our public schools, in both rural and urban settings.

Studies such as the one by Soslau and Yost (2007) have been conducted to evaluate the value of experiential education. These researchers found it to be a "viable instructional strategy to increase student learning and motivation" (p. 36). Although it takes more effort on the part of the teacher and student, the consequences are almost always positive. I found great personal success through an experiential learning process in one of my undergraduate educational courses. A professor at Keene State, with whom I still keep in contact, allowed our class to develop our own standards-based curriculum, while he acted as a mentor and offered his guidance. Since then, even though I have been fortunate to have participated in some incredible classes with professors who also promoted experiential education, I have never taken another course that I can remember in such detail. I believe it was the encouragement to try new things, experiment with innovative ideas, and utilize my personal strengths to accomplish tasks that led me to learn and comprehend more educational concepts, and eventually, make me into a more effective teacher.

With my love for risky behaviors, I have also become interested in adventure education, which is becoming a larger phenomenon every year, and is very similar to experiential education. Adventure education is hands-on learning, engaging the learner physically, emotionally, and mentally. According to Wurdinger (1997), in adventure education students are presented with a hands-on situation that they must react to immediately. The teacher encourages exploration, risk taking, play, and team cooperation (Jensen, 2000). Usually, adventure education "utilizes specific risk-taking behaviors, such as ropes courses and mountaineering, to foster personal growth" (Wurdinger, 1997, p. xi). Exciting explorations are used to enhance education, build moral character, and develop a willingness to take risks (Wurdinger, 1997). As I understand it, adventure education is not going to help students pass standardized tests, but it does wonders to increase their motivation to learn and to act cooperatively. As students are experiencing lessons, they are utilizing problem-solving skills and actively participating in their learning: they are developing more strategies to tackle hands-on activities instead of specific scholastic concepts. These skills affect their behavior in the classroom; a more engrossed and contented student is often a more engaged and comprehending student. This process of learning is different from the typical lecture and discussion techniques used in many classrooms, where the learner is expected to listen and participate, and then apply the concept at a later time (Wurdinger, 1997). What better ways are there than experiential and adventure education to demonstrate science concepts? Many science lessons involve nature in one way or another, and getting students to participate in field trips, outdoor experiential activities, and independent work outside the classroom will only help to increase their desire to learn about and love science.

Leadership Activity

Until I, a young woman, become a legitimate member within a school community, my budding leadership potential will not have a place to grow. My undergraduate experiences in a teacher preparation program and my serving as a leader on a community search and rescue team are like roots for my professional development. In order for these roots to grow and thrive in a classroom and school community, I need to be planted in an educational community in which I am a legitimate member, a classroom leader and a colleague. Based on what I have learned from these experiences, I recommend that you be unrelenting in the way that you take risks to pursue your dream job, which in rural areas can also be a hard-sought after job.

The job search process requires filling out job applications, looking in local classified ads, and applying for all teaching jobs for which you are qualified. Do not be afraid to apply for all teaching jobs for which you are qualified. If nothing else, an interview process is beneficial in giving you confidence and the ability to articulate your teaching/learning philosophy and strengths. If not offered a teaching position for the first year of one's career, you can continue to work in whatever capacity of schooling available, waiting and searching for another opportunity to show itself. Never give up. Believe that your persistence enables you to teach in the near future. Be determined that one day you will see yourself in a classroom!

In summary, the leadership activity in which I am currently engaged, and would suggest for you to take, is self-reflection. Focus on your personal qualities and the ways in which you take risks, because it is risk taking that can show and take you to the perfect opportunities. For a person attempting to enter the teaching profession, completing job applications, interviewing, and searching for the perfect opportunity is also a leadership activity.

Leadership Response

As a female leader in a rural community, I take my job as an educator seriously. My outgoing nature has produced strong leadership qualities, and I have utilized these traits in my pursuit of a full-time, permanent teaching job. My commitment to promoting education is intense, and I am not about to give up my dream. I have wanted to teach since I can remember, and limited openings in the town that I love are not going to stop me any time soon. I have become active in the Gunnison Watershed District, attending board meetings and interacting with administrators in an attempt to get name recognition. When I completed my student teaching and accepted the registrar job, I understood that I could make more money elsewhere. However more money would not have kept me in the school near the students and staff whom I had grown to respect. It would not have brought me one step closer to teaching in this community. In effect, many school and community members know me now, and compliment my drive to become a teacher.

I envision that the leadership response aspect of this chapter is the actual journey, and the sharing of my journey and self-reflection, orally and in writing, with others. I wish I had more to share about the several dimensions of teaching science, but with only weeks of experience as a part-time science educator, I feel I can give more effective advice in other arenas.

Becoming and remaining a risk taker has been the most beneficial experience of my career thus far. I have been able to maintain a connection to the school where I currently want to be, and have gained incredible knowledge by looking for jobs and participating in interviews. Interviews can be a little frightening, especially for a first-time teacher. I have learned to take deep breaths, over-prepare myself for possible questions. To prepare, I conduct research about the job, including acquiring a copy of the textbook if there is one, and if possible, speaking with current teachers. For example, for one job I actually called the man who taught the class before the gentlemen who was leaving the school, and asked for information about the curriculum and standards. One advantage/disadvantage in a small, rural town is that everyone knows each other. I use the fact that everyone knows each other to my advantage by being more comfortable and asking questions in an interview situation. Applying for several jobs and working hard in the one that I have has allowed for my name recognition in the community and has created a good reputation for myself.

It is important to know that rural schools and districts have their own politics. Because I am a first-time teacher, fairly new to the area, and only hold an undergraduate degree, I have been rejected more times than not. The schools in my geographic location want the highest quality teachers, just like anyone else, and this often means being able to boast that a certain majority hold a master's degree in the subject that they teach. I have not let this get me down, and even apply for jobs that require a master's degree.

I do not mean to discourage you from following your dreams, if they are in a small town. I dispense this advice in the hope that you too, will take risks, think outside the box, and never give up. Teaching can certainly be wonderfully rewarding in a rural town. Teaching is an incredible profession.

References

Arcidiacono, C., Procentese, F., & Napoli, I. (2007). Youth, community belonging, planning and power. *Journal of Community and Applied Social Psychology, 17*, 280–295.

Dewey, J. (1897). My pedagogic creed. *The School Journal, LIV*(3), 77–80.

Druin, G., Owens, T., & Owen S. (1986). Experiential education: A search for common roots. In R. Kraft & J. Kielsmeier (Eds.), *Experiential education and the schools* (2nd ed., pp. 50–55). Boulder, CO: Association for Experiential Education.

Haraway, D. (1988). Situated knowledges: The science question in feminism and the privilege of partial perspective. *Feminist Studies, 14*(3), 575–599.

Holland, J. (2007). Feathers of seduction. *National Geographic, 212*(1), 82–101.

Jensen, E. (2000). *Learning with the body in mind.* California: Corwin Press.

Joplin, L. (1995). On defining experiential education. In K. Warren, M. Sakois, & J. Hunt (Eds.), *The theory of experiential education.* (pp. 15–22). Dubuque, IA: Kendall/Hunt Publishing Company.

Kahneman, D., & Lovallo, D. (1993). Timid choices and bold forecasts: A cognitive perspective on risk taking. *Management Science, 39*(1), 17–31.

Kolb, D. A. (1984). *Experiential learning: Experience as the source of learning and development.* New Jersey: FT Press.

Maslow, A. H. (1943). A theory of human motivation. *Psychological Review, 50*, 370–396.

Miller, P. H. (1989). *Theories of adolescent development.* San Diego, CA: Academic Press.

Soslau, E. G., & Yost, D. S. (2007). Urban service-learning: An authentic teaching strategy to deliver a standards-driven curriculum[Electronic Version]. *Journal of Experiential Education, 30*(1), 36–53.

Wallach, M., & Kogan, N. (1964). *Risk taking: A study in cognition and personality.* Westport, CT: Greenwood Press.

Wurdinger, S. (1997). *Philosophical issues in adventure education* (3rd ed.). Dubuque, IA: Kendall/Hunt Publishing Company.

Young, R. (1991). *Risk-taking in learning, K-3.* Washington, DC: NEA Professional Library, National Education Association.

Zuckerman, M. (1979). *Beyond the optimal level of arousal.* London: Erlbaum.

Chapter 16
Epilogue: Seeds in Want of Your Watering to Germinate, Grow, Flower, Fruit, and Seed

Katherine C. Wieseman and Molly H. Weinburgh

> *Story – the abundance of it, and the lack of it – shapes us. Story – the abundance of it, and the lack of it – gives us place, lineage, history, a sense of self. Story – the abundance of it, and the lack of it breaks us into pieces, shatters our understanding, and give it back over and over again, the story different every time. Story – the abundance of it, and the lack of it – connects us with the world and outlines our relationship with everything.*
>
> *(Baldwin, 2005, p. 3)*

This book is a physical/tangible record of the knowledge and wisdom developed by and told about women in K-16 science education communities who have been guided and nourished by others, as well as acted as guides and nourished others in journeys of leadership. The stories that authors have told about their experiences are their gifts to us as readers. These stories connect us to their meaning making of leadership events and issues (Baldwin, 2005). They activate our memories, influence the relationships we have, and contribute to how we edit our own leadership histories. We, Katherine and Molly, hope that the authors' courageous steps to expose their innermost tensions, dilemmas, and problems regarding leadership issues offer guidance and inspiration as you move through your own leadership cycle. We believe that any one can lead, regardless of position, title, years of experience, and age. We also believe that we each have the responsibility to provide some leadership and direction for a shared endeavor.

As a reader, you have almost completed your journey through this book. We invite you to witness our leadership as coeditors of this book project through experiencing a behind-the-scenes story of the unfolding and shaping of this book. Listen to our story as the seeds are planted, begin to sprout, and finally flower. The purpose of the epilogue is closure from our perspectives as editors, and in it you will encounter an interweaving of our individual and collective narratives of leadership

K.C. Wieseman (✉)
Independent Educational Consultant, Formerly Western State College of Colorado, Gunnison, CO, USA
e-mail: kwieseman@gmail.com

K.C. Wieseman, M.H. Weinburgh (eds.), *Women's Experiences in Leadership in K-16 Science Education Communities, Becoming and Being*, DOI 10.1007/978-90-481-2239-4_16,

with a synthesis of similarities and differences across the chapters. Coediting the book has been a complex story of organizing, coordinating, writing, editing, mentoring, and leading each other as well as contributing authors; and being mentored and led by contributing authors and reviewers. We, like the chapter authors, are women who are becoming and being leaders in a science education community. We have chosen narrative – story – as our medium because of its power to promote the nature of leadership expressed in the book, its honoring of the relational aspects of the writing process generating the book and its promise to guide and teach (Patterson & hart-Wasekeesikaw, 1994). This story has been purposefully composed to maintain academic integrity whilst being edited for readability. For example, you will encounter adherence to APA citation rules while we share a stylized version of video chat conversations.

We have crafted the epilogue to reflect the overarching metaphor of the book, an angiosperm's life cycle. We begin with a section entitled *Germinating and Growing* in which we relate the meanings of leadership we brought to the book project. As we shared the seeds of our personal constructs with each other, we germinated and nurtured a relationship and grew in our early work as leaders of the book project. In the section *Transplanting and Growing* we introduce you to our coeditors' leadership experience and share dilemmas that we faced at different times during the book writing process. The section *Flowering and Seed Forming* reveals our synthesis of the book's content and an interpretation of similarities and differences among chapters. The epilogue ends in *New Seeds*, a section that reminds us as readers and leaders that we are each a new seed because one important dimension of leadership is that it is organic, dynamic, constant, and continuous potential.

Germinating and Growing: Personal Meanings of Leadership

We bring personal constructs of leadership to our work and service as coeditors. The questions guiding us during the early stages of co-leading a writing collaborative are, What meanings about leadership do we bring to this work? What might happen to these personal meanings? Working together gives us many opportunities to ponder these personal constructs in light of our evolving understanding of the leadership literature. Most importantly, it means being the leaders we talk and write about.

Katherine's Story

My deliberate and purposeful ponderings about the meaning of leadership and leading began during the 2007 summer holiday as a high-speed train transported me from a longtime German friend to French friends in Paris. Outside the train window the rolling hills and farmlands appeared tranquil whilst inside I was deeply pensive as I held my open journal in my lap. What in my story of leadership can be helpful for others and thus needs to be told? Picking up my pen, I scribed the following journal entry that at the time of finalizing this epilogue still rings true.

Summer 2007: To lead means to inspire! Birthing and nurturing personal and collective peace and interdependent harmony grown through moments of congruence and conflict! In other periods of my life I would have told you that leadership and leading meant to: (1) organize people, things, events so as to put some decision into action or get certain tasks done; (2) show by example in the classroom; or (3) walk one's talk not just talk a talk. Whilst simultaneously caring about the unfolding project or task, I was most dedicated to its destination – the product or outcome rather than the journey of interaction and relationship leading to the destination.

We live in a world of more rapid, increased, and even fractured interactions and events, technological sophistication, as well as separation from Earth's natural rhythms, all of which present new and different challenges for leadership and leaders. A tangible and invisible world of interaction and relationship at a very deep level is what now matters to me. What has assumed prime significance for me are the values of connectedness, interdependence, growth and understanding between people; the use of intuition; purposeful story-telling to guide and teach; and dreams as keys unlocking the pathway of one's journey, all of which are aspects of mentoring by traditional native elders according to Patterson and hart-Wasekeesikaw (1994). Secondary is the tangible and physical destination of leadership, for example this book as a product of our collaborative. Physical products of leadership in my current life might take form as a plan for team teaching, a resource handbook of contributions from and compiled for science faculty, an article, a conference presentation, or teaching as a doorway for healing. They might also be the visions, designs, structures, and specifics of educational programs. I have come to this intellectual, emotional, and spiritual space as a result of experiences in my professional and personal life. Perhaps to others my current focus on interaction and relationship is also a product, but of a different form.

How have I come to this leadership perspective? Through taking many little steps in a lifelong journey of intellect, emotion, and spirit; a study of my vulnerabilities and self; and, diverse professional and personal experiences related to my resistances, significant relationships, and a spiritual path in which I am learning to walk in two realities.

Certain professional changes that I vehemently resisted and hoped to control became key stepping stones in my leadership journey. One of these stepping stones was in my second year of teaching middle school science in Albuquerque, New Mexico, when I was put on administrative transfer to a different school due to student enrollment. Science colleagues and I made fancy arrangements to keep me at the school, but the principal would not budge in his decision. This forced transfer introduced me to Sheldon McGuire, the principal at my new school. His caring and supportive mentoring positioned me in leadership roles I never would have experienced at my first school, namely department chair, school level leadership team, mentor teacher for student teachers, and graduate student completing a master's degree in educational foundations.

People have been like cracks in my universe letting the light of leadership through. Sheldon was a huge crack in my universe. Caring acts of support and encouragement by others after him showed me unequivocally who my mentors were. Even shunning acts of judgment by others have taught me, helping me develop inner fortitude and courage. I completed my Ph.D. at the University of Georgia, a prestigious science education program in the U.S.A. I was groomed for a career in research institutions. I knew this. In the eyes of many I was expected to pursue this pathway. However, I accepted a position at a teaching rather than research institution, because as I entered the last year of my doctoral studies I knew I had to follow my heart and lead from a different vantage point. As Ronald Heifetz said about shifting consciousness regarding how an individual relates to a traditional authority – leadership equation,

> You have to learn to stomach chaos and confusion if you're going to be leading people in the midst of conflicting values, who are facing hard challenges and engaging in all sorts of avoidance behavior. You're going to have to develop a stomach for that and for not being provided with certainty (Parks, 2005, p. 37).

A third major factor in my leadership journey is a spiritual journey I am undertaking, one which involves learning to walk in two realities. These realities are (1) a physical world of the senses and understanding natural world phenomena from a Western scientific perspective, and (2) an invisible world of connectedness with dream time and spirit. I am learning to set myself aside in ways I have never known to be in service of a mystery and realness of dream time which I know has a healing aspect. These mysteries I do not completely, mentally and rationally comprehend, nor do I need or want to.

When authors submitted their chapter proposals for this book I was struck by their potential generalizeability. Celebrating the forced administrative transfer early in my career in the teaching profession was an example of Naisbitt's (2006) wisdom of viewing times of change as times of opportunity and new possibility, and making lemonade from lemons, like those experiences of Barbara Spector and Patty Simpson. Memories of my relationships with Sheldon, colleagues at Antwerp International School in Belgium, and science education faculty at the University of Georgia affirmed the importance of access into a community of practice; the influence of power relations, and socio-cultural contexts and practices; and the need for mentoring, about which Kathy Davis and Pamela Fraser-Abder write. As a global citizen whose work has been immersed in different dimensions of education, I know that my world view as a person and leader fuels my search for community who genuinely celebrate cultural diversity and the worth of each unique individual in the group. These aspects are points of focus in the chapters by Cathy Yeotis, Beth Shiner Klein, and Starlin Weaver; Hedy Moscovici and Irene Osisioma; and Tamara Nelson. Roxanne Miller's chapter about linkages between marriage and professional decision making invited me to remember my partnership during my doctoral study which dissolved because I sacrificed balance, openness, and courage for insecurity, secrecy, and an international grant research opportunity. Like Bambi Bailey, I situated myself in a teaching institution. Like Claudette Giscombe I find myself dealing with the ramifications of thinking differently (about learning, teaching, researching, and mentoring) from colleagues in my institution. Different thinking and ways of being and interacting were/are not necessarily rewarded at a small and geographically isolated institution with a more traditional leadership culture, and my recent endeavor to enact egalitarian leadership was problematic, a point about which Meta Van Sickle and Diane Cudahy write. Finally, the chapters by Robin Bingham, Lisa Pike, Wendy Files, and Sara Lamar, while quite different in focus and purpose, bring a deep smile to my countenance. These chapters situate a portion of the book's content in rural U.S.A., and particularly Wendy's and Sara's chapters remind us that the seeds and germination of leadership begin with entry into the teaching profession.

Why have I told you this as my story of leadership? Max De Pree's (2004) explanation resonates with me.

> Leadership is an art, something that is learned over time... more tribal than scientific, more a weaving of relationships than an amassing of information. (p. 3). The first responsibility of a leader is to define reality. The last is to say thank you. In between the two, the leader must become a servant and a debtor. (p. 11)

Like the authors, in sharing my story I too engage in risk taking to expose my learning and define the reality of leadership as I am experiencing it. I illuminate my weaving of relationships and serve as one of the many tribal storytellers writing this book.

I wonder what leadership means to Molly and look forward to discovering her personal ponderings. How does she see herself as a leader? What does leadership mean to her?

Molly's Story

Unlike Katherine, I was not on summer holiday. Instead I was in the United States and not in contact with Katherine. I was preparing to move from my temporary space to the new facility of the College of Education. I wondered what Katherine was thinking. What kind of deep thought was Katherine having as I tried to keep my head above water while teaching concepts of erosion in a summer school program for 8, 9, and 10 year old English Language Learners, working with graduate students, and planning events sponsored by the Andrews Institute of Mathematics, Science, and Technology Education, of which I am the director? The arrival of an email from Katherine left me smiling, and I said to myself, "Katherine is pushing me to stop my day-to-day activities long enough to think. I need conversation so I will ask a colleague to go with me for coffee and conversation." Afterward, wearing a pensive expression my fingers flew across the keyboard and my foot swung, as I typed into my informal journal that I now share to reveal my inner ponderings.

Summer 2007: Fran, Sherrie, and I spent several hours in dialogue about leadership. I have been reading books and articles on leadership – a topic I would not have ordinarily investigated. As I read and discuss thisproject with others at Texas

Christian University, I find that I become more convinced that this book is needed – a guide as well as an affirmation. Because the authors use stories to describe themselves as leaders, the book will be easy to read and offer practical suggestions in the form of leadership activities as well as be grounded in theory.

In my journal I wondered, what are my thoughts and what are the thoughts of others? In thinking about leadership, I first remember traditional ideas of leadership and the power relationships that are seen as a part of leadership. I recall the boasting statement of Glendower in Shakespeare's (n.d.) *Henry IV, Part I*, as he talks to Hotspur about his leadership ability. Glendower brags about his power as a leader, "I can call the spirits from the vasty deep" (Act III, scene 1, line 55). You can hear the emphasis on I – I can do this! What power and leadership he appears to have from this statement! However, Hotspur's reply "Why so can I or so can any man; But will they come when you do call them?" (Act II, scene 1, line 56), makes me think about the importance of the followers in leadership (Warner, 2007). Hotspur questions leadership, wanting to know if anyone will come when Glendower calls, and whether the spirits of the deep trust him enough to come. Hotspur's question is still relevant today and is being asked of and by me because being a leader is more than just calling forth the spirits with a loud and direct voice and having someone follow blindly. It is more than accomplishing a task (which I equate with bringing the spirits out of the deep), demanding of others, or getting results. It seems to me that leadership is about influencing others through shared understanding and healthy relationships, not forced compliance.

As I think about the proposed chapters that I have been reading, I reflect on how I got here, to this place at this time. Much like Lisa Pike, I did not start out wanting to be a leader. Like Meta Van Sickle, I still do not think of myself as a leader. I am Molly, a friend to, an advocate for, and a traveler in science education. The stories of leadership told by Pamela Fraser-Abder, Catherine Yeotis, Beth Shiner Klein, and Starlin Weaver helped me understand my own journey. Their descriptions of mentoring and being mentored within a community of science educators spoke to me. The words I heard were, having a community in which I feel comfortable and a mentor whom I trust, are necessary elements for my own growth as a leader. For me, mentoring and a sense of community are important and help to define who I am. Barbara Spector and Patty Simpson offered the idea of making a situation positive and valuable, whatever it was – make lemonade from lemons. Superimposed, were the stories told by Hedy and Irene about their leadership work in professional development in which I found a subtext – they often made lemonade from lemons. These stories resonated with me because I too do a great deal of work in professional development with teachers. I too appreciate seeing the positive and valuable in any situation. Even if I first question whether a situation is beneficial or not, I walk away seeing a benefit. Even though I have only mentioned a few chapters, each chapter touched me and seems to have the potential to touch a reader and to influence future leaders in our community.

My story is not unique – I began my career as a secondary biology teacher in the school in which I student taught. Like many of the authors and maybe readers, I have many (sixteen, to be exact) years of teaching experience at the secondary level. What may be different is that my school system required all teachers to begin a master's degree prior to being offered a third year contract. I missed the deadline for the state university in the area, but, as fate would have it, was admitted into Emory University. I truly believe that I would not have started a master's if it were not required by the school system, and as they say, that is the rest of the story. This one event really was a watershed in my life and career. The master's led into the doctorate and to my joining the community of science educators that I now hold as my dearest colleagues and supporters.

Here in Texas as I finish my journal entry, I know that I need to talk with Katherine. When she returns from Europe and I have moved, I want to ask her, "How can we encourage and honor the different voices of authors and external reviewers, and ours as editors while maintaining coherence and cohesiveness in the book? How can we weave the reflective, spiritual self into the leadership stories without appearing to lower academic status? How can we best encourage authors to think deeply and help them realize their own potentials?"

Transplanting and Growing: Relational Leadership in Book Writing

As autumn of 2007 arrives, physically hundreds of miles apart in our respective local settings of Fort Worth, Texas, and Gunnison, Colorado, we connect on a weekly basis using web cameras and video chat software. The video chats include nuts-and-bolts decisions (what will get done by whom and when) as well as deep intellectual discussion of leadership, the meanings of words and phrases, and our roles and feelings as leaders in this project. Additionally, we grapple with dilemmas regarding authors' intentions and chapter content and external reviewers' perceptions and feedback.

Familiar with web-based interactive video conferencing, Katherine says, "I really like it that we can see each other's facial expressions and body language as we make leadership decisions and share innermost thoughts and feelings. It is a great way to mediate our weekly meetings in cyberspace. We need and have serious and focused moments. We need and have light and laughing moments. Video chatting is enabling us to easily do this – in real time and in the moment, together."

Molly, still new to these technologies responds, "I'm still struggling with the video chatting, but you seem to be able to troubleshoot for us when needed, and it is working well to help us move forward." Then, head tilted to one side with an open expression and laughing, Molly shifts the conversation to their multidimensional responsibilities of generating the actual book. "I remember the worry I experienced through the process of writing the book proposal and sending it to the ASTE publication committee. We were successful in that endeavor. Now we are immersed in editing and writing the book, and it too will come together."

Katherine adds, "I am glad that we are working together. I so appreciate the opportunity to ask your advice as we navigate the waters of communication and feedback with authors and external reviewers. You have many gifts, of which diplomacy with words is one, which make our collaboration so pleasurable. I feel we complement each other well!"

Molly affirms, "I agree; we make a good team. We are being yet still becoming leaders in this book endeavor."

Katherine, still surprised at Molly's earlier reaction to the invitation to serve as coeditor, inquires, "Why did you say you were honored to coedit the book?"

Molly responds, "Because you could have asked any number of people, but you asked me. I am honored because I understand the need to collect, analyze and share women's experiences in leadership; to serve ASTE and WISE in this way; and to offer the gift of a book back to WISE. It [the book] is giving back to people who have given so much. This scholarship becomes a gift to upcoming generations that need to learn and be bolstered by the gifts. Giving back and giving forward."

Such humility thinks Katherine as she says, "Molly, I certainly see you as a leader because of the administrative roles and positions which you have had in science education."

Interrupting Katherine Molly exclaims, "But I do not see myself as a leader because of these formal leadership positions. Your name certainly comes to mind when we talk about leaders in WISE because, without formal title or position, you stepped forward when we needed someone to lead a collaborative co-facilitating conference sessions. You organized the logistics of WISE annual receptions; you make sure it works."

In the pause that follows, both think about the implications of perceptions: ideas of leadership associated with roles, positions, and/or titles versus leadership as activity that everyone does and can do (Astin & Leland, 1991). That a leader can be any person, "regardless of formal position, who serves as an effective social change agent" (Astin & Astin, 2000, p. 2) opposes the typical hierarchical and individualistic model in higher education: hierarchical in that "authority and power is assumed to be proportional to one's position in the administrative pecking order"(Ibid, p. 5), and individualistic in that "'leaders' among the faculty tend to be those who have gained the most professional status and recognition" (Ibid, p. 5).

Molly declares, "I see you as the true inspiration and energy behind the book. You are the leader. I am coming to know that your open, outgoing personality is only the outer layer to your being a very complex and intriguing person – one whose deep introspective nature and intense sense of values lie at the core of all you do. I am also noticing that your inner core wrestles with ideas and works out conflicts in compassionate, equitable ways. This was evident when you told me that you lead with the aim to facilitate respect, connectedness, caring, growth, and transformation between people and between people and the planet; and that you want to deepen understanding for teaching and learning as well as interactions and relationships. You are almost mystical. On the other hand, I am frank and earth bound, which explains the differences in how we have written our leadership stories."

Katherine mulls, "Yes, but like some of the authors whose chapter drafts we are reading, I struggle with conflicting beliefs and feelings. The same event leading to a sense of self-worth and value, respect, and gratitude for all the others and elders who have helped prepare me also brings forth insecurity and lack of confidence. Another dilemma of mine is wondering whether others see this book project as an opportunity for healing (balance and harmony in life, head, and heart), as we contribute individual chapter expressions, interact amongst each other as authors, leaving printed words as a finished product for generations to find, discover, make sense of, and interpret."

Molly questions, "Did you not say Christina Baldwin wrote about the power of story to heal in *Storycatcher: Making Sense of Our Lives through the Power and Practice of Story*? I see the authors doing this; they are opening and exposing themselves to their cores. We are seeing into their living room and I am so much the better for having read these women's stories."

After the video chat and in the time period before the next video chat, we are independently reflective. In Texas Molly stares at the book she should be reading while walking on the treadmill and realizes that she has not read a word. The noise of the gym fades as she thinks, "When Katherine asked me to coedit the book I told her yes. I could not have said anything else knowing that she had begun the journey by organizing symposia and long discussions with WISE members, but I am not totally comfortable. Katherine is a visionary and I am nuts-and-bolts. A good team but. . ."

Meanwhile in Colorado, Katherine alternately gazes out her office window at the mountains south of town and at her computer screen. Outside the world is seemingly tranquil. Inside her mind flutters and she hears herself saying, "We need to be so task-oriented, coordinating the logistics of whom is responsible for what, when, and how. How nice it would be to not have to issue reminders to everyone involved, even ourselves! The reality though is that we are so busy with our immediate workplace lives that these cues are needed."

Molly opens the next video chat with continued discussion about a kaleidoscope of meanings for leadership. "Whenever we use common words like the word leadership, I suspect there are more ways to understand this word than anticipated. We had something in mind when we chose this word as the focus of the book. Our dilemma now is trying to acknowledge the multifacetedness in understandings while, simultaneously, trying to be true to the authors' own definitions. We do not want to devalue others' definitions. It is such a balancing act allowing for individuality and moving a product toward conformity!"

Katherine muses about their actions as editors, "Leadership does have some broad guidelines, but how can we and external reviewers dictate an exact way chapters should be? The chapters have to parallel and model the very leadership that the authors are writing about. As editors we need to allow for and honor authors' presentations of the theoretical framework as it exists in their lives. Some authors live in the world of literature, research and theoretical perspectives; others do not. Some authors separate the story and theoretical perspectives associated with the story; others weave them together. Some tell their own story; others ask a coauthor to tell it. We need to allow for diversity in chapter development. We also need the publisher to understand and be aligned with our way of thinking – a natural flowing from and of the author, rather than just obedience to expectations to conform to a uniform framework. At the same time I say this, our and external reviewers' expectations regarding uniformity and scholarly writing are coming to the forefront."

Molly elaborates, "For example, external reviewers are seemingly counseling against a colloquial approach for a more formal approach in chapter writing. We need to remember that a potentially international audience is not likely to understand the rhythm of American colloquialism. However, we want the author's voice and way of communicating to emerge yet not sound solely academic."

Katherine expands, "Some external reviewers are commenting that the detail about accomplishments in some stories seems excessive or distracting. Should the stories show how the author's leadership journey was furthered through lessons learned from dilemmas and accomplishments, rather than appear a biography of accomplishments?"

Molly responds, "Yes. And another point – like some external reviewers, I am having trouble with the leadership activities and leadership response aspects in some of the chapters. We have talked about the purpose of the leadership activities being concrete and practical that others can do, not a continuation or reiteration of how the author was a leader. We have talked about the purpose of the leadership response being the author's opportunity to offer a meta-reflective commentary on the leadership topic, issue, and activity shared in the chapter. It is greatly problematic if outside reviewers, or we, have to struggle to find an explicit connection to leadership through these sections of a chapter. If we cannot find the connection, how will readers?"

Katherine concurs, "You are right. We really do have some broad framework for leadership within which we are operating. What ideas in the literature on leadership most resonate with you in the context of what authors are writing about as leadership and in relation to your personal meaning of leadership?"

After pausing to think about the reading she has been doing, Molly says, "Even though it is very current, Spillane's (2006) focus on how to envision a goal and successfully accomplish it in the corporate world through distributed leadership is the shadow of what I am looking for rather than the essence. De Pree's (2004) ideas on servant leadership, however, really resonate with me. They appear to be very concerned with the spiritual aspects of being a leader and the human aspect of leadership – working with people in ways that allow them to be their best and encouraging contrary opinions. De Pree wrote that good leaders abandon themselves to the strengths of others, something I keep seeing in the authors with whom we are working. I particularly like his idea of a leader being a servant and a debtor rather than a boss. Even more so, leadership should not be mechanical."

Katherine challenges, "Why does our academic side keep telling us to omit the word 'spirit' when it keeps surfacing as an aspect of leadership? Why not admit that for us and in this book connection between leadership and spirit is important? Why not put forth the question that the literature on personal qualities of leaders seems to skirt around what many people consider aspects of spirituality? Why not think about these qualities and spirituality as threads within one braid?"

We both chuckle, and the chat shifts to another topic. By this time in our coeditorship roles we have coauthored the introduction of the book, and as such experienced an external review process. Although not a one-sided blind external review as authors underwent, we too received feedback from multiple individuals that left us in a bit of a quandary because feedback from reviewers revealed contradictory perspectives and because the nature of some feedback seemed counter to our understanding of the vision of the book. This let us better understand and empathize with the intellectual and emotional processes authors might have experienced as they received and responded to reviewers' critique.

Although the chats from the summer of 2007 to 2008 often have long periods of intellectually stimulating discussions, we also have times when we figure out how to manage working with authors, external reviewers, publisher, and each other. One of us says to the other, "Oh, no! We only have 15 more minutes scheduled for this chat. What do we need to do for next time? What jobs do you want to have primary responsibility for? Did you get our logistical notes recorded in our video notes file?"

Book writing carries on to completion.

Flowering and Seed Forming: Synthesis of Similarities and Differences

Summer again has arrived and all the chapters but this epilogue are inserted in our master book file ready to send to the publisher. We are now engaged in a leadership discourse of a different nature, one focused on synthesizing the book content and interpreting it for similarities and differences between chapters.

Molly muses aloud, "What do the chapters have in common? What do they tell us collectively? What makes each chapter unique? What do they tell us individually? As a result of authors' and external reviewers' perspectives, regardless whether a leadership issue appears similar or different across chapters, what do we know now about leadership in different contexts and different fields? Even more importantly, so what? Why does this knowledge matter?"

Katherine has no immediate answer to these meta-analytical and philosophical questions. Molly continues aloud, "To me, most obvious are two cycles: (1) established leaders encouraging, nurturing, and mentoring the emergence of new leaders, and (2) broad scale leadership in K-16 science education having roots in classroom leadership. An example of the first cycle is seen in Chapter 2 as Catherine Yeotis invited Starlin Weaver and Beth Shiner Klein to join in chapter coauthorship. Established leaders mentoring and nurturing new leaders is seen in other chapters as well. An example of the second cycle is seen in the stories by Wendy Files and Sara Lamar found in Chapters 14 and 15. These authors are situated in the classroom setting, the educational context in which many of the authors of the book and perhaps science educators in general began their leadership journeys. As young leaders, Wendy and Sara are not situated in higher education yet their leadership is evident within their local context. They may or may not continue their formal education and work in academia, but they will continue to lead through their actions. These young women see leadership as a personal responsibility as well as collaborative, transformative, situational, and self-reflective. Their chapters also remind us that leadership is not necessarily connected to formal leadership positions."

Molly pauses and waits for Katherine to chime in, but Katherine continues to silently study Molly's web camera image on her computer screen. Finally, she wonders aloud, "The pervasiveness of leadership cycles is a powerful way to frame our synthesis, interpretation of chapter similarities and differences, and current understanding about leadership. Another dominant idea, perhaps not a cyclic phenomenon, is that of community. At least three-quarters of the chapters fostered leadership couched in community. The community feeds and enriches the author's leadership and the author recognizes that her leadership is taking place within a community. The literature about women (Belenky, Clinchy, Goldberger, & Tarule, 1986; Fletcher, Jordan, & Miller, 2000; Tannen, 1990) intuitively realizes that relationships are important. The literature about relationships in leadership (Graham, 1991; Uhl-Bien, 2006; Uhl-Bien, Marion, & McKelvey, 2007; Wheatley, 2006) and women in leadership (Astin & Leland, 1991; Burke & McKeen, 1990; Rhode, 2003) suggest that open supportive relationships and mentoring are necessary for 21st-century leaders. Supportive relationships imply some kind of community; therefore it is no surprise that women foster leadership in a community context and through mentoring. In earlier video chats you shared that the book's message is greater than the sum of its parts. What do you mean by this phrase?"

Molly responds, "Several chapters are grounded in the same theoretical perspective (for example, transformational leadership) but have differently enacted the theory. Enactment of the same theoretical perspective in different ways builds a richer understanding about women's leadership than any one chapter alone or any combination of chapters. Thereby, the book as a whole has greater meaning than its opening, separate sixteen chapters, and closing."

By the next video chat each of us has tackled the task of synthesizing and interpreting for similarities and differences between chapters. The first words of this chat made apparent that one of us had used inductive reasoning (writing the similarities and differences first in order to synthesize) and the other had used intuitive reasoning which to others might appear to be deductive reasoning (writing a synthesis first and then similarities and differences for confirmation). The product was the same but our pathways to the product were diametrically opposite.

Molly states, "It is clear to me that we are using different thinking styles to arrive at a synthesis of the book's content. Is there a right, or even best, way to come to a synthesis of the authors' messages to the readership? More importantly, if we use different thinking styles to lead in different ways, can we come to the same product through different pathways? Do leadership and intellectual/thinking style correlate?"

Katherine responds, "A question about relationship between leadership and intellectual or thinking style is interesting to explore. I wonder what implications such a relationship might have, and whether any relationship between leadership and thinking style, and outcome (or product) even matter. It seems to me that what is more important than association between leadership and thinking style are two other considerations: (1) leadership results in the transformation of something, and (2) the influence of the political and economic environment in which we are working might often be flying below our conscious radar. Our leadership experiences are taking place in a climate characterized by norms and values rewarding logic and reasoning rather than intuition. Perhaps this political and economic climate also promotes deductive over inductive presentation of information in a written product by leaders (for example a grant performance report written by a principal investigator or grant director, or a research study written by a research scientist), regardless of thinking style to get to the product.

Molly proclaims, "Regardless of the thinking style I used to develop my synthesis, I am drawn by how much of self each author has revealed, exposing doubts and worries as well as accomplishments. They draw upon their understanding of leadership to influence and negotiate with others to enact projects. They enact leadership responsibilities which are often associated with positions and titles. Showing their human self – their public (social) and/or private (mental and reflective) activity – does not diminish their ability to lead within their given context, and this context always has a socio-cultural dimension to it. I think that overall the leadership written about in the chapters of this book promotes a relational and caring attitude in the sense of caring as dedication, commitment, sensitivity toward feelings, and perceptions of others, and deep knowledge of self, whether the focus of the caring is on attainment of a position or title, a project's completion and success, and/or the relationships between the individuals involved. The fact that both caring and community coexist seems more important than if one is the first seed.

These leaders are not unchallenged, but rather they take risks and respond to challenges with reflection, often seeking the wisdom of a wiser, more knowledgeable other. Leadership as seen by these women involves being able to accept leadership responsibility even when it is not sought. They take a risk even when it is scary. They listen to an inner voice and act upon it, even while they might question if they are legitimate members of the discourse community or even if others do not agree. They seek alternate pathways when barriers present themselves. They work within a community even if this decision slows down a process."

Katherine explains, "Molly, how similar our syntheses are! From almost a thousand miles apart and using different approaches, we have intersected at this synthesis. I would like to return to something you just stated about the authors relating to understandings from enacting leadership responsibilities which are often associated with positions and titles. I noticed that different views about a leadership-position/title equation emerged across the chapters. A typical and historic stereotypic image and publicly accepted perception of a leader is that leadership is bequeathed by title and position. I was not surprised to encounter this view expressed in the chapters. The literature (Astin & Astin, 2000; Astin & Leland, 1991; Northouse, 2004; Shaffer, 2000; Uhl-Bien et al., 2007) had led me to anticipate that authors would equate leadership with positions/titles. Neither was I surprised by the view that authors held that anyone could lead regardless of position and title. More recent leadership literature (Sciame-Giesecke, personal communication, May 7, 2008; Uhl-Bien, 2006) espouses and promotes this idea. However, I was intrigued by the nuances of a third position, one highlighting a deep personal tension involving the historic and typical image of leadership and a contemporary emerging nonpositional image of leadership. It seems to me that in the end, one author resolved her tension by leaning more toward the leadership equated with position/title view. Another author privately lived in the anyone can lead view, and as a leader publicly tried to step away from the title and position ascribed to her. A third author resolved her tension by acting from the position that anyone can lead. Another aspect of my intrigue was that, on our first analysis for similarities we did not notice the nuances of this tension between these two views about which some authors wrote. Conducting a deeper analytical study led us to these subtle variations, thus pointing out the significance of a multidimensional approach to examining information."

Molly sighs, "Katherine, we have read these chapters many more times than a reader will, to find such differences. Do you think that a reader will notice these positions that we agree are evident from the authors' writings? Do you think the existence of positions even matters?"

Katherine forcefully reacts, "Of course the existence of different positions matters! The reader is likely to find him/herself reflected in one of the positions. This connection is important because we all seek community in some way, and finding others like ourselves provides a sense of belongingness and legitimacy in community. Are we not hoping that every reader will find at least one chapter that speaks to him/her and moves him/her forward in their leadership journey? Are we not hoping that readers feel differently, perhaps stronger and validated after reading the book? Their validation may stem from their alignment of book content. Maybe a reader senses from reading that it is okay to be different. After all, some of the authors described leadership as a journey of coming to terms with being different."

Molly pensively states, "Talking about being different reminds me of some differences between chapters. The uniquenesses tempt us to present all of them as a laundry list but..."

Katherine interrupts, "Yes, but instead let us only focus on a few of the differences because of our perception of their significance."

Molly reciprocates with, "Exactly. First, few of the authors explicitly consistently featured family relationships in their stories of leadership. Those that did wrote about goal setting and pursuit, sense of belongingness, and influence of faith in shaping all of the author's leadership style. Those that did not might have mentioned that dreams for their children, spousal reaction, or spousal career had some bearing on a decision to take a position or charted a direction at a career intersection. I find it telling when an author does not disclose personal life factors that we know have been critical influences in the author's life and their identities as leaders. I find it telling that all of us are in families in some way. Families shape us and impact us, as well as are impacted by our career decision making. Yet families often have been omitted from the leadership stories that the authors believe have the greatest value."

Katherine continues, "Second, the nature of one's socio-cultural context determines the culture in which leadership flowers and fruits. Every leadership story, activity, and response is set in a socio-cultural context, but individual perceptions illuminate the extent of cultural continuity or discontinuity with the broader community. Whether each individual's capital is expressed, heard, and acknowledged, or whether individuals truly agree about how a decision is taken (100 percent consensus or majority democracy as written by authors) are two important and different dimensions of the socio-cultural context of leadership."

Molly carries on, "Third, all the authors have passion and passion has driven them to lead. However, their passion is different in that it has taken them into different arenas, such as a teaching college, professional development of teachers, a research agenda that deviates from the norm, success in the two things held most dear – a desire to stay in academia and marriage, and the K-12 classroom."

Katherine finishes with, "Finally, I would include that leadership can be from individuals working fulltime in K-12 worlds, and this leadership as authors have written about it, often prioritizes K-12 student achievement and teacher accountability for student learning in ways that none of the authors situated fulltime in academia have featured."

After a moment of pause, Molly wonders, "So after reading and analyzing any work, I ask myself so what? As a reader you too may be asking yourself, so what? We hope that this book has made you think about yourself and leadership, and the intersection of the two. Some of the dilemmas shared in the book have prevailed over time and lead us to ask many questions."

Katherine jumps in quickly with, "Are the leadership dilemmas in the book eternal in the private realm for each of us as individuals? Are they eternal in a public realm? Will there always be a dilemma of women working in men's spaces? Will a time come when women's and men's capital in the workplace are regarded horizontally rather than vertically, so difference ceases to be a dilemma for women? Will a time come when the trailing spouse phenomenon is perceived by self and others without surprise or judgment, thus becoming less of a public dilemma for the partners involved?"

Molly chimes in, "Is society changing so that at least some of the leadership dynamics and challenges expressed in the book become obsolete? Can friction be removed for those who are educated in research 1/research extensive institutions, but choose a different workplace setting from which to lead? Can leadership be any voice calling forth – quiet, boisterous, spiritual, practical – with others participating and following? Can leadership exist without a group? Can leadership exist without interaction and exchange? Are great ideas alone sufficient basis to lead?"

Katherine smiles, "Such philosophical questions! They are part of our so what discussion and have emerged from our thinking about the chapters. Like any good investigation, we are left with more questions than answers."

Molly responds, "Furthermore we think that the essence of leadership will probably remain eternal, although the veneer may change."

New Seeds: Anyone Can Lead

The leadership cycle continues. We see emerging women leaders around us in all levels of K-16 science education. Our work with other women who are leaders (being) and who feel that they are still growing as leaders (becoming) has cemented our convictions that women are powerful leaders. We witness and celebrate women teaching and mentoring next generations within communities of learners, professional development, organizations, classrooms, scholarly research, and informal gatherings.

The leadership cycle continues. We see readers selecting this book because they are interested in the topic of leadership. As a reader, you now have multiple examples of women's ways of leading. By reading and interpreting the book, and doing the leadership activities within the chapters, we hope that you are furthering your being and becoming a leader. You are a seed. In fertile ground your leadership germinates, grows, and can flower and yield fruit to distribute even more seeds.

We believe that anyone can lead and that each of us should lead.

References

Astin, A. A. & Astin, H. S. (2000). *Leadership reconsidered: Engaging higher education in social change.* Battle Creek, MI: W.K. Kellogg Foundation.

Astin, H. S., & Leland, C. (1991). *Women of influence, women of vision: A cross-generational study of leaders and social change.* San Francisco: Jossey-Bass Publishers.

Baldwin, C. (2005). *Storycatcher: Making sense of our lives through the power and practice of story.* Novato, CA: New World Library.

Belenky, M. F., Clinchy, B. M., Goldberger, N. R., & Tarule, J. M. (1986). *Women's ways of knowing: The development of self, voice and mind.* New York: Basic Books.

Burke, R. J., & McKeen, C. A. (1990). Mentoring in organizations: Implications for women. *Journal of Business Ethics, 9*(4–5), 317–332.

De Pree, M. (2004). *Leadership is an art.* New York: Dell Publishing.

Fletcher, J. K., Jordan, J. V., & Miller, J. B. (2000). Women and the workplace: Applications of a psychodynamic theory. *The American Journal of Psychoanalysis, 60*, 243–261.

Graham, J. W. (1991). Servant-leadership in organizations: Inspirational and moral. *Leadership Quarterly, 2*(2), 105–119.

Naisbitt, J. (2006). *Mind set! Reset your thinking and see the future.* New York: HarperCollins.

Northouse, P. G. (2004).*Leadership: Theory and practice* (3rd ed.). Thousand Oaks, CA: Sage.

Parks, S. D. (2005). *Leadership can be taught: A bold approach for a complex world.* Boston, MA: Harvard Business School Press.

Patterson, B., & hart-Wasekeesikaw, F. (1994). Mentoring women in higher education: Lessons from the elders. *College Teaching, 94*(2), 72–78.

Rhode, D. (2003). *The differences 'differences' make: Women and leadership.* Palato: Stanford University Press.

Shakespeare, W. (n.d.). *Henry IV, Part I.* Retrieved November 20, 2007, from http://www.shakespeare-online.com/plays/1kh4_3_1.html

Shaffer, J. C. (2000). *The leadership solution.* New York: McGraw Hill.

Spillane, J. P. (2006). *Distributed leadership.* San Francisco: Jossey-Bass Publishers.

Tannen, D. (1990). *You just don't understand: Women and men in conversation.* New York: William Morrow.

Uhl-Bien, M. (2006). Relational Leadership Theory: Exploring the social processes of leadership and organizing. *The Leadership Quarterly, 17*(6), 654–676.

Uhl-Bien, M., Marion, R., & McKelvey, B. (2007). Complexity Leadership Theory: Shifting leadership from the industrial age to the knowledge era. *The Leadership Quarterly, 18*(4), 298–318.

Warner, N. (2007). Screening leadership through Shakespeare: Paradoxes of leader-follower relations in Henry V on film. *The Leadership Quarterly, 18*(1), 1–15.

Wheatley, M. J. (2006). *Leadership and the new science.* San Francisco: Berrett-Koehler Publishers.

Closing: The Cycle is Complete, But Can It Begin Again?

The following passages from the *Invisible Man* by Ralph Ellison (1972) have had a profound impact on my world view as a science teacher educator and researcher. An unknown African American male, the protagonist in the book stated:

> I saw him approach several people and turn away without speaking. He's lost, I thought, and he'll keep coming until he see me, then he'll ask his direction. Maybe there's an embarrassment in it if he admits he's lost .... Perhaps to lose a sense of *where* you are implies the danger of losing a sense of who you are. That must be it, I thought. ... to lose your direction is to lose your face. So here he comes to ask his direction from the lost, the invisible. Very well, I've learned to live without direction. Let him ask (p. 564).

> I'm coming out, no less invisible without it, but coming out nevertheless. I've overstayed my hibernation since there's a possibility that even an invisible man has a socially responsive role to play. Being invisible and without substance, a disembodied voice, as it were, what else could I do? What else but try to tell you what was really happening when your eyes were looking through? And it is this which frightens me: Who knows but that, on the lower frequencies, I speak for you? (p. 568)

How can someone with brown skin that must function in a dominating world of people with much less melatonin in their skin be invisible? But many are, especially the children in the schools in the United States. Many children with brown skin hibernate in the schools until they can leave school and work in a world that wishes them to continue to be invisible.

Invisibility is not a characteristic of leaders and those desiring to be leaders. I believe it is an honor and an unexpected privilege, that I – a Black female who lives in a world of oppression and domination and struggles to be a change agent to eliminate oppression and domination – am the author of the closing chapter of a book, which is a record of the knowledge and wisdom developed by and told about female leaders in K-16 science education communities. I am bound by my positionality of culture, race, gender, class, and ability. These elements of my positionality are also bound by time – when I was born, where I grew up, and my life experiences. If I tell you that I learned to function into two cultures at an early age – the African American culture and the dominant culture because my father was in the military, then what does this information tell you about me? In my early years, I went to an elementary school in Germany where I was one of the few children of color. Hence, I can ask the question: Are the authors of these chapters representative of the voices of all in science teaching or science teacher education? The membership of most of the science education teacher in education and research organizations is male and White in the United States. If the science education schooling is to change so that all students have the opportunity for quality science teaching and learning, then the leaders in public education must diversify.

In the closing of the book it is imperative that I point out that certain voices are missing from the leadership in science education and in the book. These voices are the voices of women of color who have struggled and many times been very unsuccessful in becoming leaders in science education. These voices need to be expressed and acknowledged for the cycle of leadership to begin again. The coeditors of the book did approach the others so that the voices of Asian American, Latina, and Native American and more Black females would be heard. However, many of these women may have chosen not be included for two reasons: painful memories are not what one wants to relive and re-experience and/or the exposure of self to others is fear-laden due to job security or provides others the information to become gatekeepers. Gate keeping occurs when the powerful withholds something that the less powerful values and/or wants (Fairclough, 1989). However, the powerful discourse in this book is not about controlling and constraining the contributions of the less powerful. Some of the authors do share instances of gate-keeping encounters, while others do not share their instances of gate keeping. For instance, one author shares her commitment to tell the stories of an urban middle school African American student and other female scientists of color, but she writes little about Blackness and classism as she discusses (1) White teachers' reluctance to venture in the worlds of their own urban Black students or (2) her own struggles to earn her doctorate in science education. Another author, a more senior author, relays her struggles to diversify the teacher education in her program, but provides no details of her struggle to become a leader of color in science education. Other authors do not share their own interactions with other women of color. Hopefully, the readers of this book do not overlook the silence of some women of color and, hence, contribute to invisibility of other women's experiences in the book. If some of the readers forget this silence and invisibility, then the missing discourse in this book might contribute to the reproduction of social structures in science education.

Even though the coeditors ask the authors to ground their stories in theoretical frameworks, one theoretical framework is missing – womanism. The perspective of womanism builds on and comes from women's experiences, in particular women of color (Collins, 1986, 1990; Hooks, 1984; Olesen, 1994; Tillman, 2002). Womanism is viewed as a social change philosophical orientation, a theoretical paradigm that focuses on the struggles of women of color based on racism, classism, sexism, heterosexism, and ageism to give voice to the experiences of these women and for women to act in the struggle for human freedom and social justice (deMarrais & LeCompte, 1995; Grant & Ladson-Billings, 1997; Phillips, 2006). In her seminal book *Black Feminist Thought: Knowledge, Consciousness, and the Politics of Empowerment*, Collins (1990) offered three levels on which to consider the lived experience of domination, oppression, and empowerment: personal biography level (individual); the group or community level where race-, class-, and gender-based influences intersect (social); and the systemic level of social institutions (societal). I am attempting in this closing chapter to begin the discussion of incorporating womanism into the thinking of the WISE group.

Young women of color, some of the new seeds, must be afforded the opportunities to develop their leadership knowledge and skills in the science education arena (Atwater, 2000). These arenas include science teaching, science teacher education, and science education research. WISE can serve as one of these arenas. New seeds, particularly women of color, need sponsors and mentors for them to achieve so the cycle can begin. Otherwise, the cycle will not begin again or the cycle will begin with less diversity and the science education community will forever lose the benefits of these women's voices and their work.

As I read the chapters in this book, the language used reminds me of the power of words. For how you define yourself tells others a lot about whom you are, whom you want to be, and for what you will stand in the difficult times of life. Words such as personal and social responsibility, collaboration, courage, transformation, risk taking, and cultural border crossing are the foundation of the lives of many of these female authors and leaders. Married female leaders juggle two lives, the family and the profession. Sometimes they are successful and sometimes they are unsuccessful in juggling these lives. I once had a female international doctoral student ask me if you [a female] could have it all. Since I was also young and married, I believed then that anything was possible. I answered her question with a resounding "yes." I now have lived longer and have more knowledge and understanding of the world, so now my answer would be "yes sometimes" and "no sometimes." Single females have the extra struggle of developing partnerships to meet their emotional and social needs. Developing these partnerships is sometimes difficult when you are living in a country in which you did not grow up or in a state or a region of a country without any family. Difficult times in the lives of females are when female leaders especially use their courage, collaborative skills, and sense of responsibility to take risks and cross borders to transform themselves and to help others to transform themselves, to become leaders so they can change their worlds and the worlds of others for the better. The journey of becoming a leader is not easy for males; it may be even more difficult for females. Some female leaders must be pioneers. I remember a White male administrator once saying to me, "Mary, you are a pioneer." I looked at him and said, "Being a pioneer is difficult and the road is not easy." I would now add that it is difficult for the people around you – family, colleagues, those who build barriers and put obstacles in your way, or those who sit on the sidelines not assisting you in any manner to forge that trail so eventually a road can be built. A female leader like Barbara Spector is one of those pioneers in science education who fought for the visibility and voices of women. I remember when Barbara Spector would convince females in science education to gather together over a dinner meal to discuss women's issues and provide encouragement. She extended personal invitations to the few women of color in science education in late 1980s and early 1990s. She was one of the builders of the trail so that WISE could be established.

Finally, one must remember that I am only one voice in the community of science education. If someone else had been chosen to write the closing of the book, her perspective would have been different because she would have had very different experiences. These experiences would have influenced her world view and what she would have emphasized in the book's closing. For me, I must remember that most of the female leaders in science education pursue their missions due to the students in the schools. I will not forget the words of Vivian Gordon in *Black Women, Feminism, and Black Liberation: Which Way?* quoted by C. Hudson-Weems (2006). I have altered the wordings to be more inclusive:

> To address women's issues, therefore, is not only to address the crucial needs of [all] women, it is also to address the historic primacy of [all the communities found in the world]; that is, the primacy of its children and their preparation for the responsibilities and privileges of mature personhood (p. 38).

I maintain that this journey into science education leadership is worth all females taking, and the journey will become easier with time. How much time it will take, I do not know. This is one of the challenges of WISE.

Professor of Science Education
and Adjunct Professor of Social Foundations,
University of Georgia
Spring 2008

Mary Atwater

References

Atwater, M. M. (2000). Females in science education: White is the norm and class, language, life style, and religion are non-issues. *Journal of Research in Science Teaching, 37*, 386–387.

Collins, P. H. (1986). Learning from the outsider within: The sociological significance of Black feminist thought. *Social Problems, 33*, 514–532.

Collins, P. H. (1990). *Black feminist thought: Knowledge, consciousness and the politics of empowerment.* New York: Routledge.

deMarrais, K. B., & LeCompte, M. D. (1995). *The way schools work: A sociological analysis of education.* White Plains, NY: Longman.

Ellison, R. (1972). *Invisible man.* New York: Random House.

Fairclough, N. (1989). *Language and power.* New York: Longman.

Grant, C. A., & Ladson-Billings, G. (Eds.) (1997). *Dictionary of multicultural education.* Phoenix, AZ: Oryx Press.

Hooks, B. (1984). *Feminist theory from margins to center.* Boston: South End Press.

Hudson-Weems, C. (2006). Cultural and agenda conflicts in academia: Critical issues for Africana women's studies. In L. Phillips (Ed.), *The womanist reader* (pp. 37–43). New York: Routledge.

Olesen, V. (1994). Feminism and models of qualitative research. In N. K. Denzin & Y. S. Lincoln (Eds.), *Handbook of qualitative research* (pp. 158–174). Thousand Oaks, CA: Sage Publications.

Phillips, L. (2006). Womanism: On its own. In L. Phillips (Ed.), *The womanist reader* (pp. xix–iv). New York: Routledge.

Tillman, L. C. (2002). Culturally sensitive research approaches: An African American perspective. *Educational Researcher, 31*(9), 3–12.

Author Index

Subject Index

M

N

O

Printed in the United States
149810LV00003B/5/P

9 789048 122363